SAS® 9.3 Macro Language
Reference

SAS® Documentation

The correct bibliographic citation for this manual is as follows: SAS Institute Inc. 2011. *SAS® 9.3 Macro Language: Reference*. Cary, NC: SAS Institute Inc.

SAS® 9.3 Macro Language: Reference

ISBN 978-1-60764-942-7 (electronic book)
ISBN 978-1-60764-894-9

SAS Institute Inc., SAS Campus Drive, Cary, North Carolina 27513.

ISBN 978-1-60764-942-7
1st electronic book, July 2011

ISBN 978-1-60764-894-9
1st printing, July 2011

SAS® Publishing provides a complete selection of books and electronic products to help customers use SAS software to its fullest potential. For more information about our e-books, e-learning products, CDs, and hard-copy books, visit the SAS Publishing Web site at `support.sas.com/publishing` or call 1-800-727-3228.

Contents

About This Book

Syntax Conventions for the SAS Language

Overview of Syntax Conventions for the SAS Language

SAS uses standard conventions in the documentation of syntax for SAS language elements. These conventions enable you to easily identify the components of SAS syntax. The conventions can be divided into these parts:

- syntax components
- style conventions
- references to SAS libraries and external files

Syntax Components

The components of the syntax for most language elements include a keyword and arguments. For some language elements only a keyword is necessary. For other language elements the keyword is followed by an equal sign (=).

Note: In most cases, example code in SAS documentation is written in lowercase with a monospace font. You can use uppercase, lowercase, or mixed case in the code that you write.

Style Conventions

The style conventions that are used in documenting SAS syntax include uppercase bold, uppercase, and italic:

UPPERCASE BOLD
identifies SAS keywords such as the names of functions or statements. In the following example, the keyword ERROR is written in uppercase bold:

ERROR<message>;

UPPERCASE
identifies arguments that are literals. In the following example of the CMPMODEL= system option, the literals include BOTH, CATALOG, and XML:

CMPMODEL = BOTH | CATALOG | XML

italics
identifies arguments or values that you supply. Items in italics represent user-supplied values that are either non-literal arguments or nonliteral values that are assigned to an argument.

Items in italics can also be the generic name for a list of arguments from which you can choose (for example, *attribute-list*). If more than one of an item in italics can be used, the items are expressed as *item-1, ..., item-n*.

References to SAS Libraries and External Files

Many SAS statements and other language elements refer to SAS libraries and external files. You can choose whether to make the reference through a logical name (a libref or fileref) or use the physical filename enclosed in quotation marks. If you use a logical name, you usually have a choice of using a SAS statement (LIBNAME or FILENAME) or the operating environment's control language to make the association. Several methods of referring to SAS libraries and external files are available, and some of these methods depend on your operating environment.

In the examples that use external files, SAS documentation uses the italicized phrase *file-specification*. In the examples that use SAS libraries, SAS documentation uses the italicized phrase *SAS-library*. Note that *SAS-library* is enclosed in quotation marks:

```
infile file-specification obs = 100;
libname libref 'SAS-library';
```

What's New in the SAS 9.3 Macro Language Facility

Overview

The Macro Language Facility has the following enhancements:

- new automatic macro variables that enable you to reduce the amount of text that is needed to perform common tasks
- new macro functions
- new macro statements
- new macro system options that enable you to define and redefine macros and to better control their execution

Automatic Macro Variables

SYSADDRBITS
: contains the number of bits of an address.

SYSENDIAN
: contains an indication of the byte order of the current session. The possible values are LITTLE or BIG.

SYSNOBS
: contains the number of observations read from the last data set that was closed by the previous procedure or DATA step.

SYSODSESCAPECHAR
: displays the value of the ODS ESCAPECHAR= from within the program.

SYSSIZEOFLONG
: contains the length in bytes of a long integer in the current session.

SYSSIZEOFPTR
: contains the size in bytes of a pointer.

SYSSIZEOFUNICODE
contains the length in bytes of a Unicode character in the current session.

Macro Functions

%SYSMACEXEC
indicates whether a macro is currently executing.

%SYSMACEXIST
indicates whether there is a macro definition in the WORK.SASMACR catalog.

%SYSMEXECDEPTH
returns the depth of nesting from the point of call.

%SYSMEXECNAME
returns the name of the macro executing at a nesting level.

Macro Statements

%SYSMSTORECLEAR
closes stored compiled macros and clears the SASMSTORE= library.

%SYSMACDELETE
deletes a macro definition from the WORK.SASMACR catalog.

Macro System Options

MAUTOCOMPLOC
displays in the SAS log the source location of the autocall macros when the autocall macro is compiled.

MAUTOLOCINDES
specifies whether the macro processor prepends the full pathname of the autocall source file to the description field of the catalog entry of compiled auto call macro definition in the WORK.SASMACR catalog.

MCOVERAGE
enables the generation of coverage analysis data.

MCOVERAGELOC=
specifies the location of the coverage analysis data file.

Recommended Reading

- *Carpenter's Complete Guide to the SAS Macro Language*
- *Debugging SAS Programs: A Handbook of Tools and Techniques*
- *SAS Macro Programming Made Easy*
- *Base SAS Procedures Guide*
- *SAS Language Reference: Concepts*
- *SAS Functions and CALL Routines: Reference*

For a complete list of SAS publications, go to support.sas.com/bookstore. If you have questions about which titles you need, please contact a SAS Publishing Sales Representative:

SAS Publishing Sales
SAS Campus Drive
Cary, NC 27513-2414
Phone: 1-800-727-3228
Fax: 1-919-677-8166
E-mail: sasbook@sas.com
Web address: support.sas.com/bookstore

Part 1

Understanding and Using the Macro Facility

Chapter 1

Introduction to the Macro Facility

Getting Started with the Macro Facility

This document is the macro facility language reference for SAS. It is a reference for the SAS macro language processor and defines the SAS macro language elements. This section introduces the SAS macro facility using simple examples and explanation.

The *macro facility* is a tool for extending and customizing SAS and for reducing the amount of text that you must enter to do common tasks. The macro facility enables you to assign a name to character strings or groups of SAS programming statements. You can work with the names that you created rather than with the text itself.

The SAS macro language is a string based language. It does not support the use of hexadecimal character constants.

Note: The SAS macro language does not support using hexadecimal values to specify non-printable characters.

When you use a macro facility name in a SAS program or from a command prompt, the macro facility generates SAS statements and commands as needed. The rest of SAS receives those statements and uses them in the same way it uses the ones that you enter in the standard manner.

The macro facility has two components:

macro processor
is the portion of SAS that does the work

macro language
is the syntax that you use to communicate with the macro processor

When SAS compiles program text, two delimiters trigger macro processor activity:

&name
refers to a macro variable. “Replacing Text Strings Using Macro Variables” on page 4 explains how to create a macro variable. The form *&name* is called a macro variable reference.

%name
refers to a macro. “Generating SAS Code Using Macros ” on page 5 explains how to create a macro. The form *%name* is called a macro call.

The text substitution produced by the macro processor is completed before the program text is compiled and executed. The macro facility uses statements and functions that resemble the statements and functions that you use in the DATA step. An important difference, however, is that macro language elements can enable only text substitution and are not present during program or command execution.

Note: Three SAS statements begin with a % that are not part of the macro facility. These elements are the %INCLUDE, %LIST, and %RUN statements in *SAS Statements: Reference*

The following graphic explains the syntax used in this document:

Syntax Conventions

PROC DATASETS <LIBRARY=*libref*> <MEMTYPE=(*mtype-list*)>
<DETAILS | NODETAILS> <*other-options*>;
RENAME *variable-1*=*new-name-1* <. . . *variable-n*=*new-name-n*>;

1 SAS keywords, such as statement or procedure names, appear in bold type.
2 Values that you must spell as they are given in the syntax appear in uppercase type.
3 Optional arguments appear inside angle brackets(<>).
4 Mutually exclusive choices are joined with a vertical bar(|).
5 Values that you must supply appear in italic type.
6 Argument groups that you can repeat are indicated by an ellipsis (. . .).

Replacing Text Strings Using Macro Variables

Macro variables are an efficient way of replacing text strings in SAS code. The simplest way to define a macro variable is to use the %LET statement to assign the macro variable a name (subject to standard SAS naming conventions), and a value.

```
%let city=New Orleans;
```

Now you can use the macro variable CITY in SAS statements where you would like the text **New Orleans** to appear. You refer to the variable by preceding the variable name with an ampersand (&), as in the following TITLE statement:

```
title "Data for &city";
```

The macro processor resolves the reference to the macro variable CITY:

```
title "Data for New Orleans";
```

A macro variable can be defined within a macro definition or within a statement that is outside a macro definition (called *open code*).

Note: The title is enclosed in double quotation marks. In quoted strings in open code, the macro processor resolves macro variable references within double quotation marks but not within single quotation marks.

A %LET statement in open code (outside a macro definition) creates a global macro variable that is available for use anywhere (except in DATALINES or CARDS statements) in your SAS code during the SAS session in which the variable was created. There are also *local* macro variables, which are available for use only inside the macro definition where they are created. See Scope of Macro Variables on page 43 for more information about global and local macro variables.

Macro variables are not subject to the same length limits as SAS data set variables. However, if the value that you want to assign to a macro variable contains certain special characters (for example, semicolons, quotation marks, ampersands, and percent signs) or mnemonics (for example, AND, OR, or LT), you must use a macro quoting function to mask the special characters. Otherwise, the special character or mnemonic might be misinterpreted by the macro processor. See Macro Quoting on page 80 for more information.

While macro variables are useful for simple text substitution, they cannot perform conditional operations, DO loops, and other more complex tasks. For this type of work, you must define a macro.

Generating SAS Code Using Macros

Defining Macros

Macros enable you to substitute text in a program and to do many other things. A SAS program can contain any number of macros, and you can invoke a macro any number of times in a single program.

To help you learn how to define your own macros, this section presents a few examples that you can model your own macros after. Each of these examples is fairly simple; by mixing and matching the various techniques, you can create advanced, flexible macros that are capable of performing complex tasks.

Each macro that you define has a distinct name. When choosing a name for your macro, it is recommended that you avoid a name that is a SAS language keyword or call routine name. The name that you choose is subject to the standard SAS naming conventions. (See the Base SAS language documentation for more information about SAS naming conventions.) A macro definition is placed between a %MACRO statement and a %MEND (macro end) statement, as in the following example:

%MACRO *macro-name*;

%MEND *macro-name*;

The *macro-name* specified in the %MEND statement must match the *macro-name* specified in the %MACRO statement.

Note: While specifying the *macro-name* in the %MEND statement is not required, it is recommended. It makes matching %MACRO and %MEND statements while debugging easier.

Here is an example of a simple macro definition:

```
%macro dsn;
   Newdata
%mend dsn;
```

This macro is named DSN. **Newdata** is the text of the macro. A string inside a macro is called *constant text* or *model text* because it is the model, or pattern, for the text that becomes part of your SAS program.

To call (or *invoke*) a macro, precede the name of the macro with a percent sign (%):

%macro-name

Although the call to the macro looks somewhat like a SAS statement, it does not have to end in a semicolon.

For example, here is how you might call the DSN macro:

```
title "Display of Data Set %dsn";
```

The macro processor executes the macro DSN, which substitutes the constant text in the macro into the TITLE statement:

```
title "Display of Data Set Newdata";
```

Note: The title is enclosed in double quotation marks. In quoted strings in open code, the macro processor resolves macro invocations within double quotation marks but not within single quotation marks.

The macro DSN is exactly the same as the following coding:

```
%let dsn=Newdata;

title "Display of Data Set &dsn";
```

The following code is the result:

```
title "Display of Data Set Newdata";
```

So, in this case, the macro approach does not have any advantages over the macro variable approach. However, DSN is an extremely simple macro. As you will see in later examples, macros can do much more than the macro DSN does.

Inserting Comments in Macros

All code benefits from thorough commenting, and macro code is no exception. There are two forms that you can use to add comments to your macro code.

The first form is the same as comments in SAS code, beginning with **/*** and ending with ***/**. The second form begins with a **%*** and ends with a **;**. The following program uses both types of comments:

```
%macro comment;
/* Here is the type of comment used in other SAS code. */
   %let myvar=abc;

%* Here is a macro-type comment.;
   %let myvar2=xyz;

%mend comment;
```

You can use whichever type comment that you prefer in your macro code, or use both types as in the previous example.

The asterisk-style comment (* *commentary* ;) used in SAS code is not recommended within a macro definition. While the asterisk-style will comment constant text appropriately, it will execute any macro statements contained within the comment. This form of comment is not recommended because unmatched quotation marks contained within the comment text are not ignored and can cause unpredictable results.

Macro Definition Containing Several SAS Statements

You can create macros that contain entire sections of a SAS program:

```
%macro plot;
   proc plot;
      plot income*age;
   run;
%mend plot;
```

Later in the program that you can invoke the macro:

```
data temp;
   set in.permdata;
   if age>=20;
run;

%plot

proc print;
run;
```

When these statements execute, the following program is produced:

```
data temp;
   set in.permdata;
   if age>=20;
run;

proc plot;
   plot income*age;
run;

proc print;
run;
```

Passing Information into a Macro Using Parameters

A macro variable defined in parentheses in a %MACRO statement is a *macro parameter*. Macro parameters enable you to pass information into a macro. Here is a simple example:

```
%macro plot(yvar= ,xvar= );
   proc plot;
      plot &yvar*&xvar;
   run;
%mend plot;
```

You invoke the macro by providing values for the parameters:

```
%plot(yvar=income,xvar=age)

%plot(yvar=income,xvar=yrs_educ)
```

When the macro executes, the macro processor matches the values specified in the macro call to the parameters in the macro definition. (This type of parameter is called a *keyword parameter*.)

Macro execution produces the following code:

```
proc plot;
   plot income*age;
run;

proc plot;
   plot income*yrs_educ;
run;
```

Using parameters has several advantages. First, you can write fewer %LET statements. Second, using parameters ensures that the variables never interfere with parts of your program outside the macro. Macro parameters are an example of *local* macro variables, which exist only during the execution of the macro in which they are defined.

Conditionally Generating SAS Code

By using the %IF-%THEN-%ELSE macro statements, you can conditionally generate SAS code with a macro. See the following example:

```
%macro whatstep(info=,mydata=);
   %if &info=print %then
      %do;
         proc print data=&mydata;
         run;
      %end;

   %else %if &info=report %then
      %do;
         options nodate nonumber ps=18 ls=70 fmtsearch=(sasuser);
      proc report data=&mydata nowd;
         column manager dept sales;
         where sector='se';
         format manager $mgrfmt. dept $deptfmt. sales dollar11.2;
         title 'Sales for the Southeast Sector';
      run;
   %end;
%mend whatstep;
```

In this example, the macro WHATSTEP uses keyword parameters, which are set to default null values. When you call a macro that uses keyword parameters, specify the parameter name followed by an equal sign and the value that you want to assign the parameter. Here, the macro WHATSTEP is called with INFO set to **`print`** and MYDATA set to **`grocery`**.

```
%whatstep(info=print,mydata=grocery)
```

This code produces the following statements:

```
proc print data=grocery;
run;
```

Because values in the macro processor are case sensitive, the previous program does not work if you specify **`PRINT`** instead of **`print`**. To make your macro more robust, use the %UPCASE macro function. For more information, see "%UPCASE and %QUPCASE Functions" on page 275.

For more information, see "%MACRO Statement" on page 304 and "%MEND Statement" on page 310.

More Advanced Macro Techniques

Generating Repetitive Pieces of Text Using %DO Loops

"Conditionally Generating SAS Code" on page 8 presents a %DO-%END group of statements to conditionally execute several SAS statements. To generate repetitive pieces of text, use an iterative %DO loop. For example, the following macro, NAMES, uses an iterative %DO loop to create a series of names to be used in a DATA statement:

```
%macro names(name= ,number= );
   %do n=1 %to &number;
      &name&n
   %end;
%mend names;
```

The macro NAMES creates a series of names by concatenating the value of the parameter NAME and the value of the macro variable N. You supply the stopping value for N as the value of the parameter NUMBER, as in the following DATA statement:

```
data %names(name=dsn,number=5);
```

Submitting this statement produces the following complete DATA statement:

```
data dsn1 dsn2 dsn3 dsn4 dsn5;
```

Note: You can also execute a %DO loop conditionally with %DO %WHILE and %DO %UNTIL statements. For more information, see "%DO %WHILE Statement" on page 292 and "%DO %UNTIL Statement" on page 291.

Generating a Suffix for a Macro Variable Reference

Suppose that, when you generate a numbered series of names, you always want to put the letter X between the prefix and the number. The macro NAMESX inserts an X after the prefix you supply:

```
%macro namesx(name=,number=);
   %do n=1 %to &number;
      &name.x&n
   %end;
%mend namesx;
```

The period is a delimiter at the end of the reference &NAME. The macro processor uses the delimiter to distinguish the reference &NAME followed by the letter X from the reference &NAMEX. Here is an example of calling the macro NAMESX in a DATA statement:

```
data %namesx(name=dsn,number=3);
```

Submitting this statement produces the following statement:

```
data dsnx1 dsnx2 dsnx3;
```

See Macro Variables on page 19 for more information about using a period as a delimiter in a macro variable reference.

Other Features of the Macro Language

Although subsequent sections go into far more detail on the various elements of the macro language, this section highlights some of the possibilities, with pointers to more information.

macro statements
: This section has illustrated only a few of the macro statements, such as %MACRO and %IF-%THEN. Many other macro statements exist, some of which are valid in open code, while others are valid only in macro definitions. For a complete list of macro statements, see "Macro Statements " on page 156.

macro functions
: Macro functions are functions defined by the macro facility. They process one or more arguments and produce a result. For example, the %SUBSTR function creates a substring of another string, while the %UPCASE function converts characters to uppercase. A special category of macro functions, the macro quoting functions, mask special characters so they are not misinterpreted by the macro processor.

  There are two special macro functions, %SYSFUNC and %QSYSFUNC, that provide access to SAS language functions or user-written functions generated with SAS/TOOLKIT. You can use %SYSFUNC and %QSYSFUNC with new functions in Base SAS software to obtain the values of SAS host, base, or graphics options. These functions also enable you to open and close SAS data sets, test data set attributes, or read and write to external files. Another special function is %SYSEVALF, which enables your macros to perform floating-point arithmetic.

  For a list of macro functions, see "Macro Functions " on page 158. For a discussion of the macro quoting functions, see Macro Quoting on page 80. For the syntax of calling selected Base SAS functions with %SYSFUNC, see Syntax for Selected Functions with the %SYSFUNC Function on page 367.

autocall macros
: Autocall macros are macros defined by SAS that perform common tasks, such as trimming leading or trailing blanks from a macro variable's value or returning the data type of a value. For a list of autocall macros, see "Selected Autocall Macros Provided with SAS Software" on page 168.

automatic macro variables
: Automatic macro variables are macro variables created by the macro processor. For example, SYSDATE contains the date SAS is invoked. See "Macro Language Elements" on page 155 for a list of automatic macro variables.

macro facility interfaces
: Interfaces with the macro facility provide a dynamic connection between the macro facility and other parts of SAS, such as the DATA step, SCL code, the SQL procedure, and SAS/CONNECT software. For example, you can create macro variables based on values within the DATA step using CALL SYMPUT and retrieve the value of a macro variable stored on a remote host using the %SYSRPUT macro statement. For more information about these interfaces, see Interfaces with the Macro Facility on page 101.

Chapter 2
SAS Programs and Macro Processing

SAS Programs and Macro Processing

This section describes the typical pattern that SAS follows to process a program. These concepts are helpful for understanding how the macro processor works with other parts of SAS. However, they are not required for most macro programming. They are provided so that you can understand what is going on behind the scenes.

Note: The concepts in this section present a logical representation, not a detailed physical representation, of how SAS software works.

When you submit a program, it goes to an area of memory called the *input stack*. This is true for all program and command sources: the SAS windowing environment, the SCL SUBMIT block, the SCL COMPILE command, or from batch or noninteractive sessions. The input stack shown in the following figure contains a simple SAS program that displays sales data. The first line in the program is the top of the input stack.

Figure 2.1 *Submitted Programs Are Sent to the Input Stack*

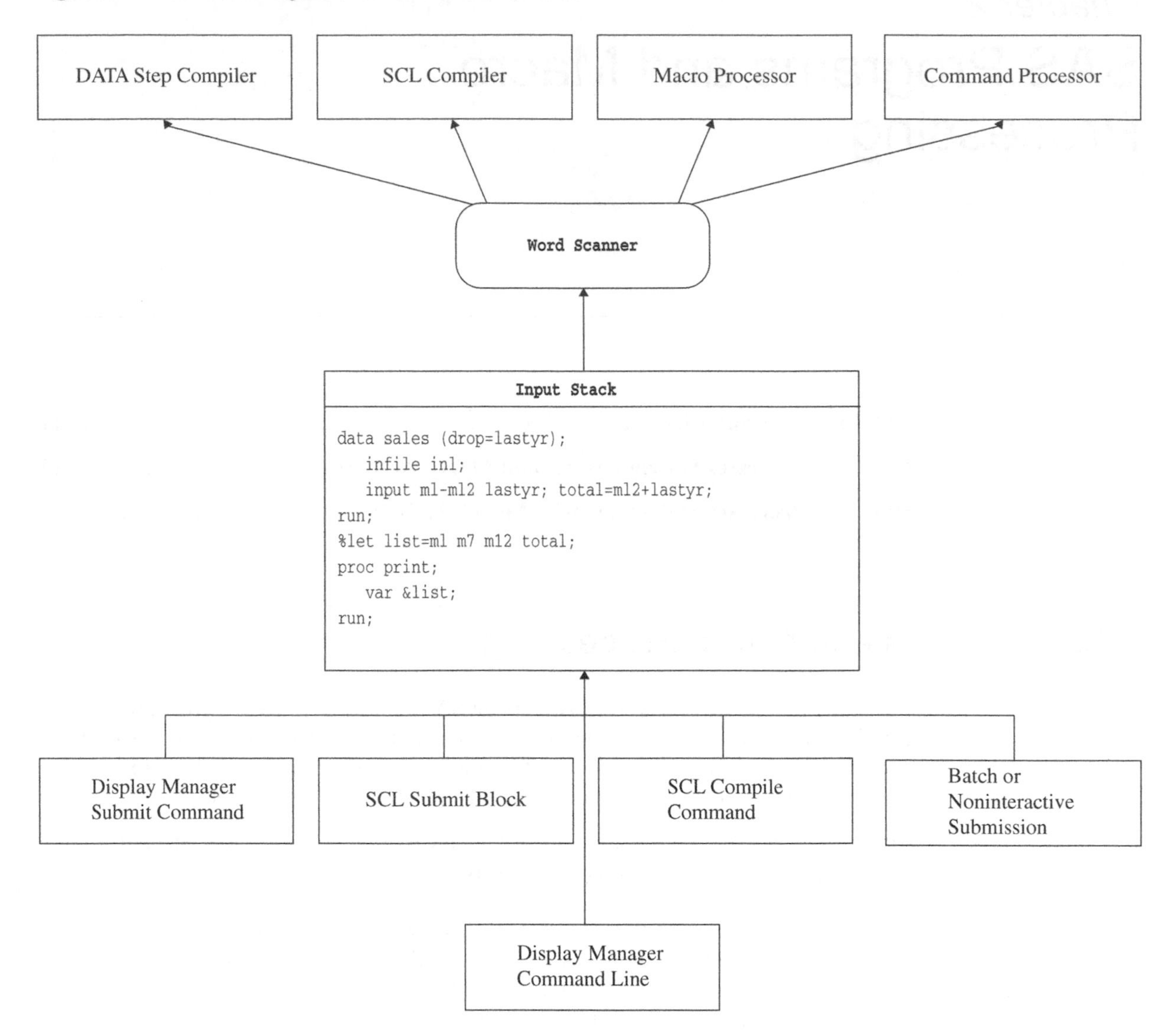

Once a program reaches the input stack, SAS transforms the stream of characters into individual tokens. These tokens are transferred to different parts of SAS for processing, such as the DATA step compiler and the macro processor. Knowing how SAS recognizes tokens and how they are transferred to different parts of SAS will help you understand how the various parts of SAS and the macro processor work together. Also how to control the timing of macro execution in your programs. The following sections show you how a simple program is tokenized and processed.

How SAS Processes Statements without Macro Activity

The process that SAS uses to extract words and symbols from the input stack is called *tokenization*. Tokenization is performed by a component of SAS called the *word scanner*, as shown in Figure 2.2 on page 13. The word scanner starts at the first character in the input stack and examines each character in turn. In doing so, the word scanner assembles the characters into tokens. There are four general types of tokens:

Literal
: a string of characters enclosed in quotation marks.

Number
: digits, date values, time values, and hexadecimal numbers.

Name
: a string of characters beginning with an underscore or letter.

Special
: any character or group of characters that have special meaning to SAS. Examples of special characters include: `* / + - ** ; $ ( ) . & % =`

Figure 2.2 *The Sample Program before Tokenization*

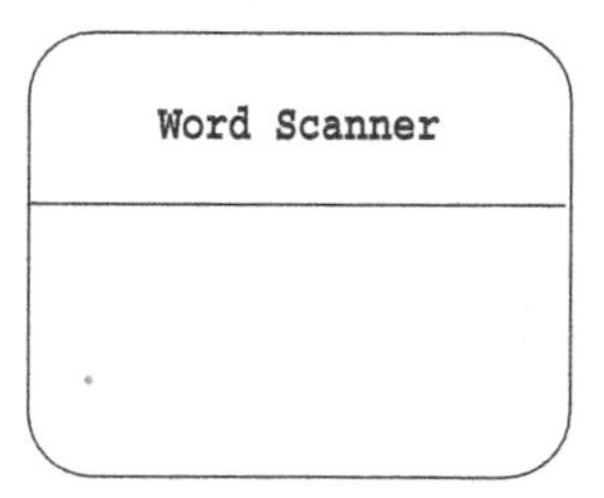

Input Stack

```
data sales (drop=lastyr);
   infile in1;
   input m1-m12 lastyr;
   total=m12+lastyr;
run;
```

The first SAS statement in the input stack in the preceding figure contains eight tokens (four names and four special characters).

```
data sales(drop=lastyr);
```

When the word scanner finds a blank or the beginning of a new token, it removes a token from the input stack and transfers it to the bottom of the queue.

In this example, when the word scanner pulls the first token from the input stack, it recognizes the token as the beginning of a DATA step. The word scanner triggers the DATA step compiler, which begins to request more tokens. The compiler pulls tokens from the top of the queue, as shown in the following figure.

Figure 2.3 *The Word Scanner Obtains Tokens*

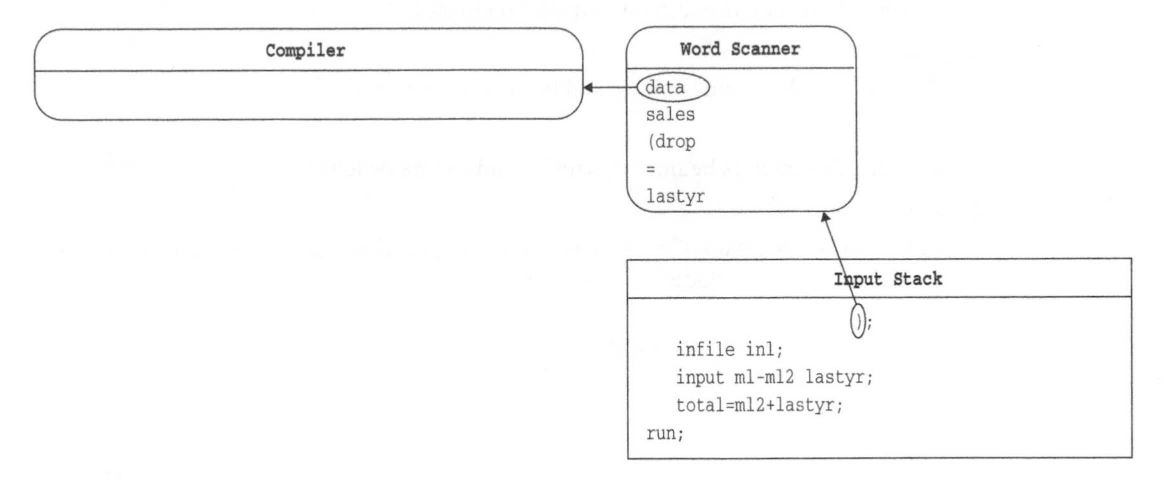

The compiler continues to pull tokens until it recognizes the end of the DATA step (in this case, the RUN statement), which is called a DATA step boundary, as shown in the following figure. When the DATA step compiler recognizes the end of a step, the step is executed, and the DATA step is complete.

Figure 2.4 *The Word Scanner Sends Tokens to the Compiler*

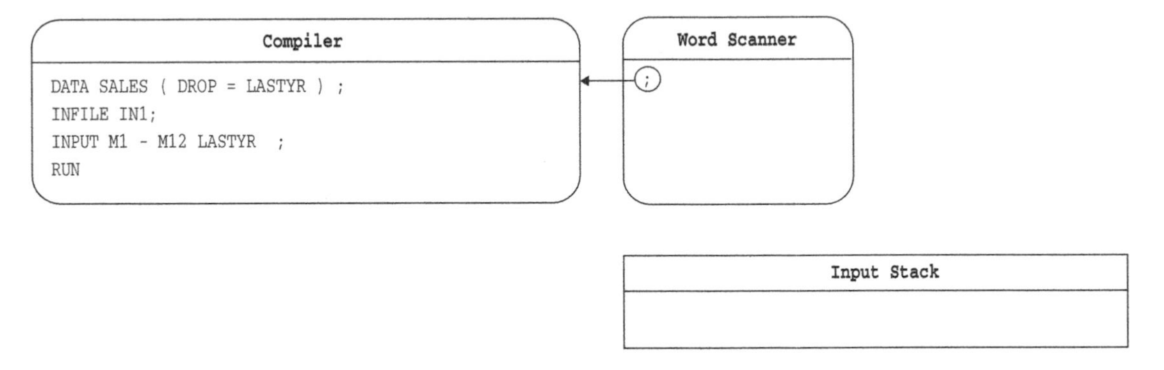

In most SAS programs with no macro processor activity, all information that the compiler receives comes from the submitted program.

How SAS Processes Statements with Macro Activity

In a program with macro activity, the macro processor can generate text that is placed on the input stack to be tokenized by the word scanner. The example in this section shows you how the macro processor creates and resolves a macro variable. To illustrate how the compiler and the macro processor work together, the following figure contains the macro processor and the macro variable symbol table. SAS creates the symbol table at the beginning of a SAS session to hold the values of automatic and global macro variables. SAS creates automatic macro variables at the beginning of a SAS session. For the sake of illustration, the symbol table is shown with only one automatic macro variable, SYSDAY.

Figure 2.5 The Macro Processor and Symbol Table

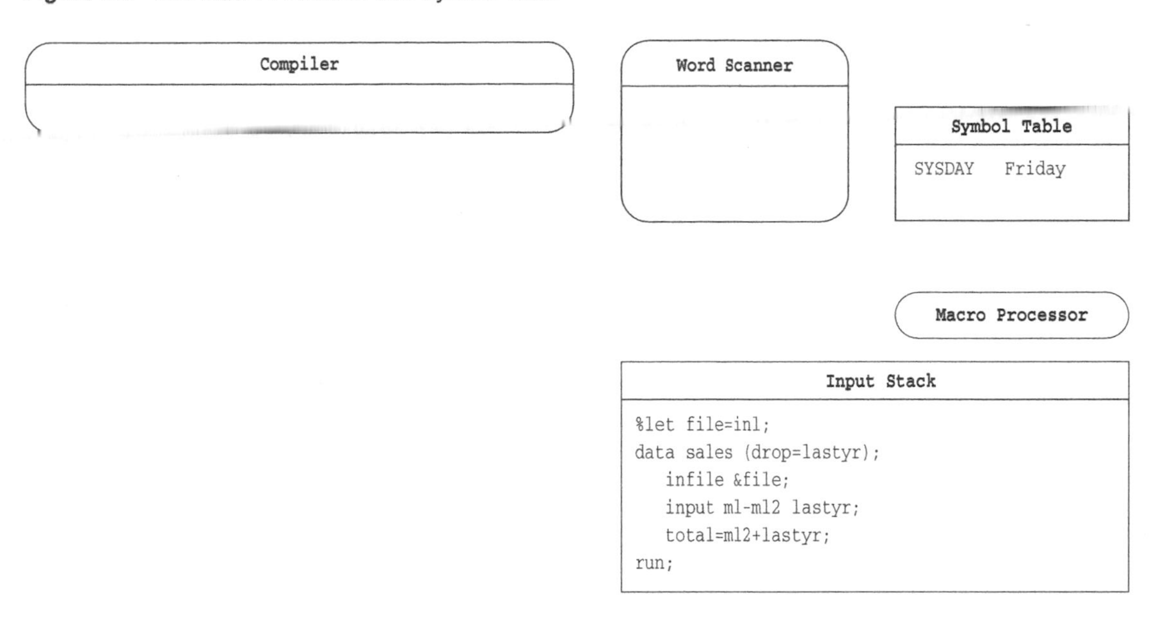

Whenever the word scanner encounters a macro trigger, it sends information to the macro processor. A macro trigger is either an ampersand (&) or percent sign (%) followed by a nonblank character. As it did in the previous example, the word scanner begins to process this program by examining the first characters in the input stack. In this case, the word scanner finds a percent sign (%) followed by a nonblank character. The word scanner recognizes this combination of characters as a potential macro language element, and triggers the macro processor to examine % and LET, as shown in the following figure.

Figure 2.6 The Macro Processor Examines LET

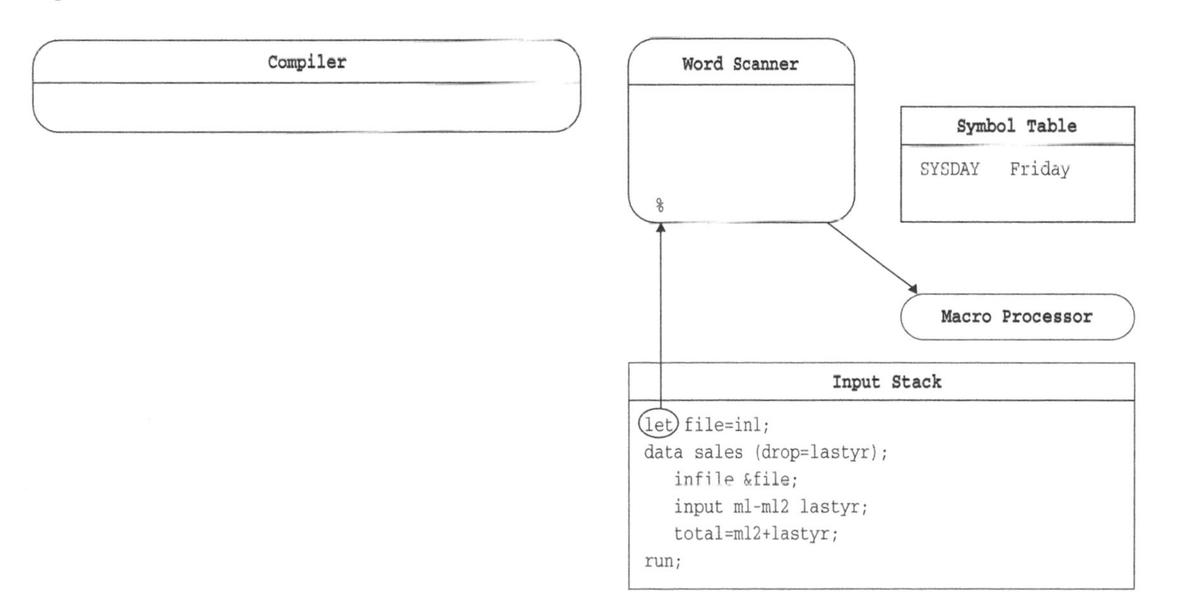

When the macro processor recognizes a macro language element, it begins to work with the word scanner. In this case, the macro processor removes the %LET statement, and writes an entry in the symbol table, as shown in the following figure.

Figure 2.7 *The Macro Processor Writes to the Symbol Table*

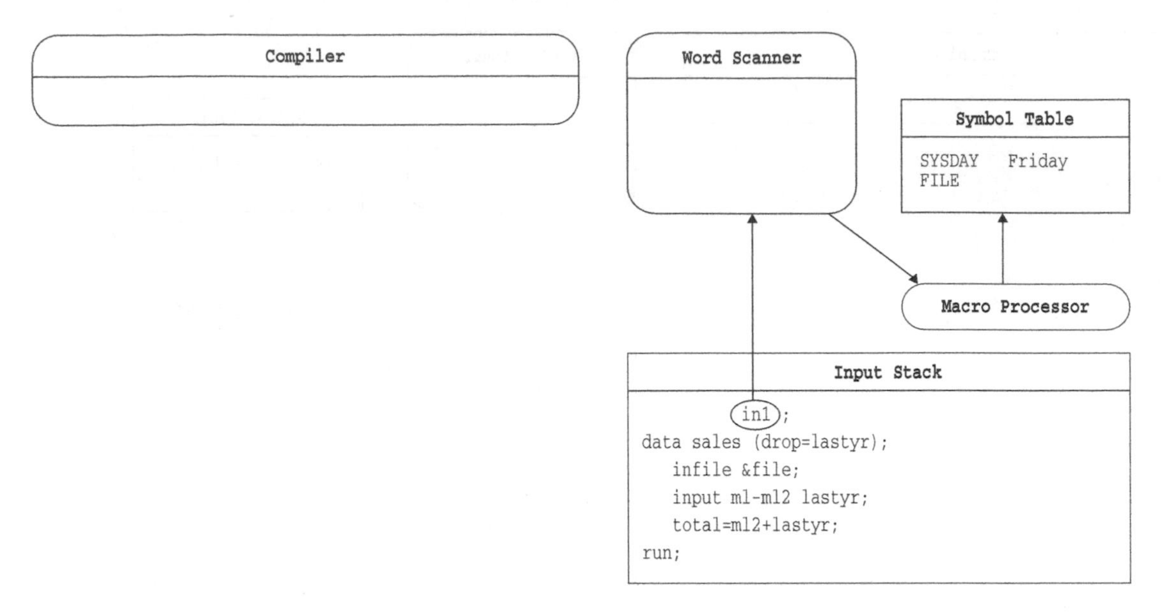

From the time the word scanner triggers the macro processor until that macro processor action is complete, the macro processor controls all activity. While the macro processor is active, no activity occurs in the word scanner or the DATA step compiler.

When the macro processor is finished, the word scanner reads the next token (the DATA keyword in this example) and sends it to the compiler. The word scanner triggers the compiler, which begins to pull tokens from the top of the queue, as shown in the following figure.

Figure 2.8 *The Word Scanner Resumes Tokenization*

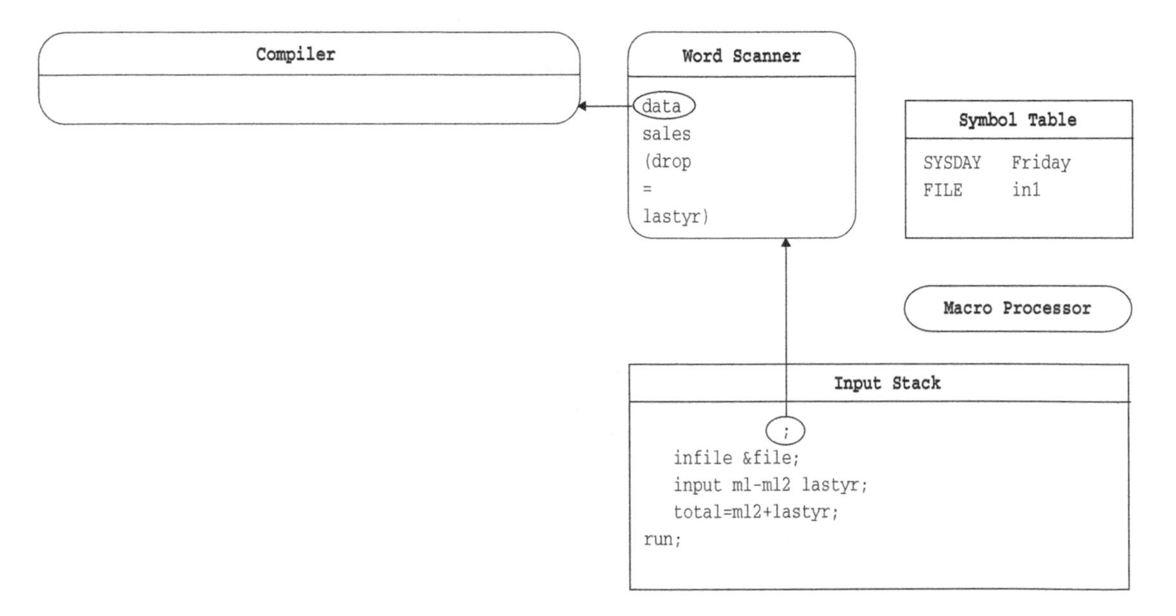

As it processes each token, SAS removes the protection that the macro quoting functions provide to mask special characters and mnemonic operators. For more information, see "Macro Quoting" on page 80.

If the word scanner finds an ampersand followed by a nonblank character in a token, it triggers the macro processor to examine the next token, as shown in the following figure.

Figure 2.9 The Macro Processor Examines &FILE

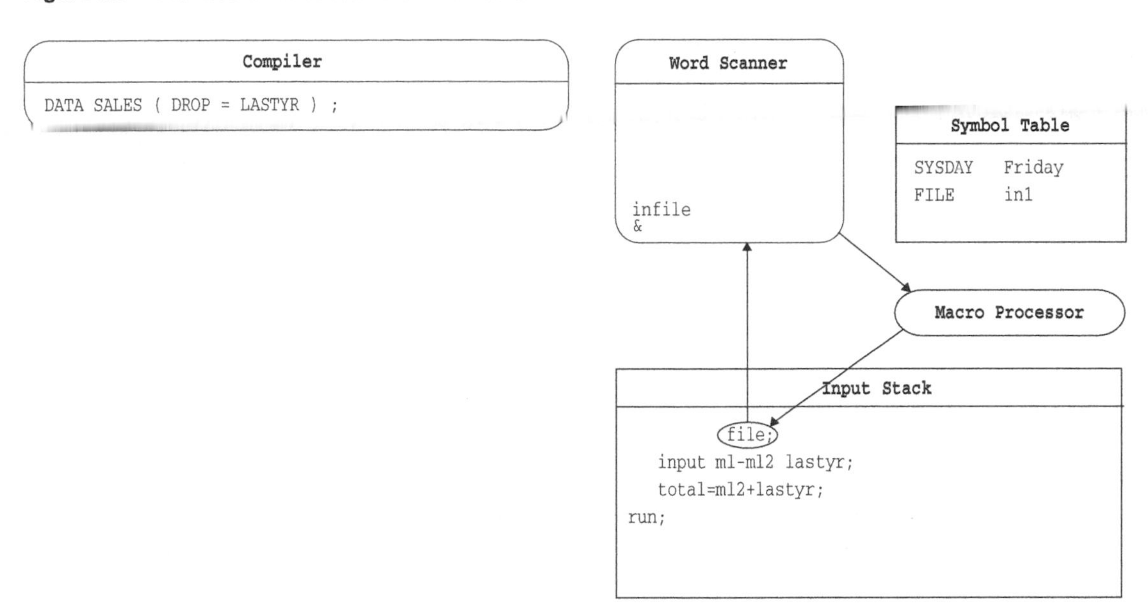

The macro processor examines the token and recognizes a macro variable that exists in the symbol table. The macro processor removes the macro variable name from the input stack and replaces it with the text from the symbol table, as shown in the following figure.

Figure 2.10 The Macro Processor Generates Text to the Input Stack

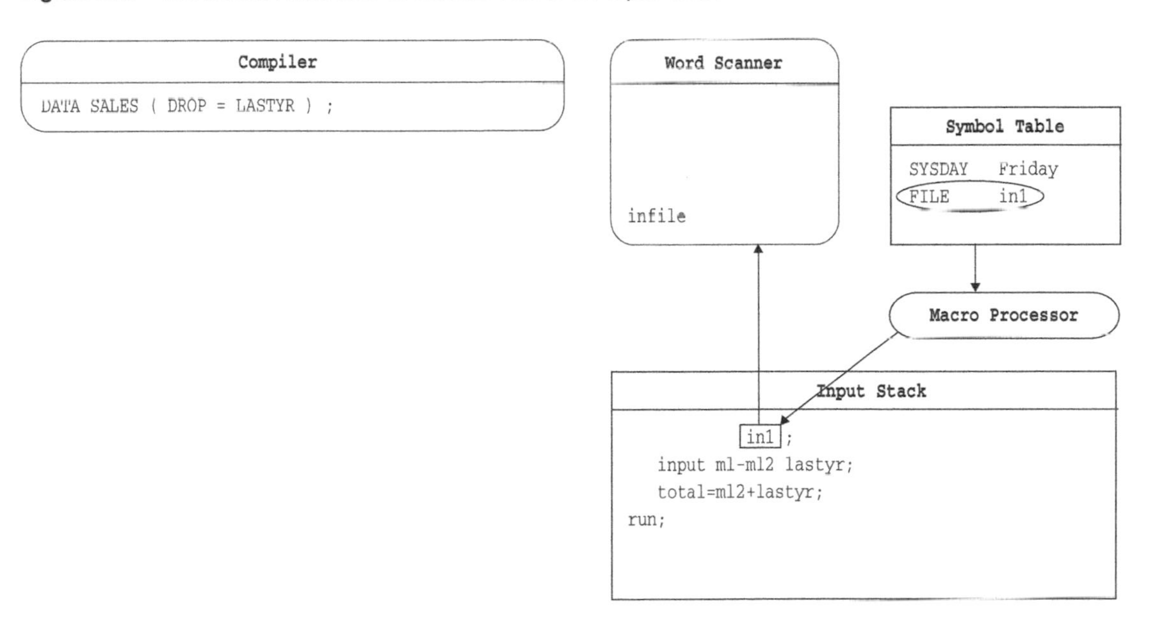

The compiler continues to request tokens, and the word scanner continues to supply them, until the entire input stack has been read as shown in the following figure.

Figure 2.11 *The Word Scanner Completes Processing*

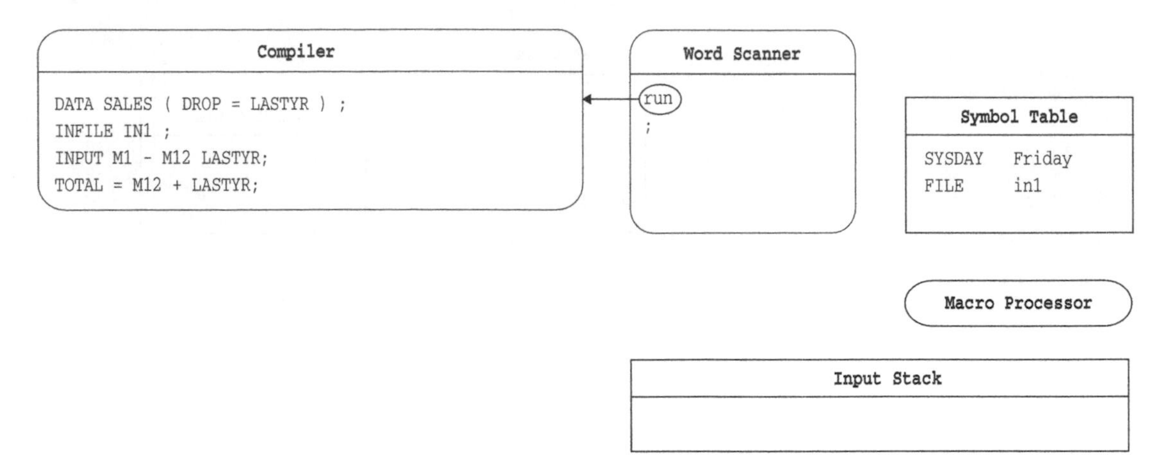

If the end of the input stack is a DATA step boundary, as it is in this example, the compiler compiles and executes the step. SAS then frees the DATA step task. Any macro variables that were created during the program remain in the symbol table. If the end of the input stack is not a step boundary, the processed statements remain in the compiler. Processing resumes when more statements are submitted to the input stack.

Chapter 3
Macro Variables

Macro Variables

Macro variables are tools that enable you to dynamically modify the text in a SAS program through symbolic substitution. You can assign large or small amounts of text to macro variables, and after that, you can use that text by simply referencing the variable that contains it.

Macro variable values have a maximum length of 65,534 characters. The length of a macro variable is determined by the text assigned to it instead of a specific length declaration. So its length varies with each value that it contains. Macro variables contain only character data. However, the macro facility has features that enable a variable to be evaluated as a number when it contains character data that can be interpreted as a number. The value of a macro variable remains constant until it is specifically changed. Macro variables are independent of SAS data set variables.

Note: Only printable characters should be assigned to macro variables. Non-printable values that are assigned to macro variables can cause unpredictable results.

Macro variables defined by macro programmers are called *user-defined macro variables*. Those defined by the macro processor are called *automatic macro variables*. You can define and use macro variables anywhere in SAS programs, except within data lines.

When a macro variable is defined, the macro processor adds it to one of the program's macro variable symbol tables. When a macro variable is defined in a statement that is outside a macro definition (called *open code*) or when the variable is created automatically by the macro processor (except SYSPBUFF), the variable is held in the global symbol table, which the macro processor creates at the beginning of a SAS session. When a macro variable is defined within a macro and is not specifically defined as global, the variable is typically held in the macro's local symbol table, which SAS creates when the macro starts executing. For more information about symbol tables, see “SAS Programs and Macro Processing” on page 11 and “Scopes of Macro Variables” on page 43.

When it is in the global symbol table, a macro variable exists for the remainder of the current SAS session. A variable in the global symbol table is called a *global macro variable*. This variable has global scope because its value is available to any part of the SAS session (except in CARDS or DATALINES statements). Other components of SAS might create global macro variables, but only those components created by the macro processor are considered *automatic macro variables*.

When it is in a local symbol table, a macro variable exists only during execution of the macro in which it is defined. A variable in a local symbol table is called a *local macro variable*. It has local scope because its value is available only while the macro is executing. “SAS Programs and Macro Processing” on page 11 contains figures that illustrate a program with a global and a local symbol table.

You can use the %PUT statement to view all macro variables available in a current SAS session. See “%PUT Statement” on page 310 and also in “Macro Facility Error Messages and Debugging” on page 120.

Macro Variables Defined by the Macro Processor

When you invoke SAS, the macro processor creates automatic macro variables that supply information related to the SAS session. Automatic variables are global except SYSPBUFF, which is local.

To use an automatic macro variable, reference it with an ampersand followed by the macro variable name (for example, &SYSJOBID). This FOOTNOTE statement contains references to the automatic macro variables SYSDAY and SYSDATE9:

```
footnote "Report for &sysday, &sysdate9";
```

If the current SAS session is invoked on December 16, 2011, macro variable resolution causes SAS to receive this statement:

```
FOOTNOTE "Report for Friday, 16DEC2011";
```

Automatic macro variables are often useful in conditional logic such as a %IF statement with actions determined by the value that is returned. For more information, see “%IF-%THEN/%ELSE Statement” on page 296.

You can assign values to automatic macro variables that have read and write status. However, you cannot assign a value to an automatic macro variable that has read-only status. The following table lists the automatic macro variables that are created by the SAS macro processor and their read and /write status.

Use %PUT _AUTOMATIC_ to view all available automatic macro variables.

There are also system-specific macro variables that are created only on a particular platform. These are documented in the host companion, and common ones are listed in "Writing Efficient and Portable Macros" on page 141. Other SAS software products also provide macro variables, which are described in the documentation for the product that uses them. Neither of these types of macro variables are considered automatic macro variables.

Table 3.1 *Automatic Macro Variables by Category*

Status	Variable	Contains
Read and Write	SYSBUFFR	Unmatched text from %INPUT
	SYSCC	The current condition code that SAS returns to your operating environment (the operating environment condition code)
	SYSCMD	Last unrecognized command from the command line of a macro window
	SYSDEVIC	Name of current graphics device
	SYSDMG	Return code that reflects an action taken on a damaged data set
	SYSDSN	Name of most recent SAS data set in two fields
	SYSFILRC	Return code set by the FILENAME statement
	SYSLAST	Name of most recent SAS data set in one field
	SYSLCKRC	Return code set by the LOCK statement
	SYSLIBRC	Return code set by the LIBNAME statement
	SYSLOGAPPLNAME	Value of the LOGAPPLNAME option
	SYSMSG	Message for display in macro window
	SYSPARM	Value specified with the SYSPARM= system option
	SYSPBUFF	Text of macro parameter values
	SYSRC	Various system-related return codes
Read-only	SYSADDRBITS	The number of bits of an address
	SYSCHARWIDTH	The character width value
	SYSDATE	The character value representing the date a SAS job or session began executing (two-digit year)

Status	Variable	Contains
	SYSDATE9	The character value representing the date a SAS job or session began executing (four-digit year)
	SYSDAY	Day of week SAS job or session began executing
	SYSENCODING	Name of the SAS session encoding
	SYSENDIAN	An indication of the byte order of the current session
	SYSENV	Foreground or background indicator
	SYSERR	Return code set by SAS procedures and the DATA step
	SYSERRORTEXT	Text of the last error message formatted for display on the SAS log
	SYSHOSTNAME	Host name of the operating environment
	SYSINDEX	Number of macros that have begun execution during this session
	SYSINFO	Return code information
	SYSJOBID	Name of current batch job or user ID (varies by host environment)
	SYSMACRONAME	Name of current executing macro
	SYSMENV	Current macro execution environment
	SYSNCPU	The current number of processors that SAS might use in computation
	SYSNOBS	The number of observations read from the last data set
	SYSODSESCAPECHAR	The value of the ODS ESCAPECHAR= from within the program
	SYSODSPATH	The value of the PATH variable in the Output Delivery System (ODS)
	SYSPROCESSID	The process ID of the current SAS process
	SYSPROCESSNAME	The process name of the current SAS process
	SYSPROCNAME	Name of current procedure being processed
	SYSSCP	The abbreviation of an operating system
	SYSSCPL	The name of an operating system

Status	Variable	Contains
	SYSSITE	The number assigned to your site
	SYSSIZEOFLONG	The length in bytes of a long integer in the current session
	SYSSIZEOFPTR	The size in bytes of a pointer
	SYSSIZEOFUNICODE	The length in bytes of a Unicode character in the current session
	SYSSTARTID	The ID generated from the last STARTSAS statement
	SYSSTARTNAME	The process name generated from the last STARTSAS statement
	SYSTCPIPHOSTNAME	Host names of the local and remote operating environments when multiple TCP/IP stacks are supported
	SYSTIME	The character value of the time a SAS job or session began executing
	SYSUSERID	The user ID or login of the current SAS process
	SYSVER	Release or version number of SAS software executing
	SYSVLONG	Release number and maintenance level of SAS software with a 2-digit year
	SYSVLONG4	Release number and maintenance level of SAS software with a 4-digit year
	SYSWARNINGTEXT	Text of the last warning message formatted for display on the SAS log

Macro Variables Defined by Users

Overview for Defining Macro Variables

You can create your own macro variables, change their values, and define their scope. You can define a macro variable within a macro, and you can also specifically define it as a global variable, by defining it with the %GLOBAL statement. Macro variable names must start with a letter or an underscore and can be followed by letters or digits. You can assign any name to a macro variable as long as the name is not a reserved word. The prefixes AF, DMS, SQL, and SYS are not recommended because they are frequently used in SAS software when creating macro variables. Thus, using one of these prefixes can cause a name conflict with macro variables created by SAS software.

For a complete list of reserved words in the macro language, see " Reserved Words in the Macro Facility" on page 363. If you assign a macro variable name that is not valid, an error message is printed in the SAS log.

You can use %PUT _ALL_ to view all user-created macro variables.

Creating Macro Variables and Assigning Values

The simplest way to create and assign a value to a macro variable is to use the macro program statement %LET:

```
%let dsname=Newdata;
```

DSNAME is the name of the macro variable. **Newdata** is the value of the macro variable DSNAME. The value of a macro variable is simply a string of characters. The characters can include any letters, numbers, or printable symbols found on your keyboard, and blanks between characters. The case of letters is preserved in a macro variable value. Some characters, such as unmatched quotation marks, require special treatment, which is described later.

If a macro variable already exists, a value assigned to it replaces its current value. If a macro variable or its value contains macro triggers (% or &), the trigger is evaluated before the value is assigned. In the following example, **&name** is resolved to **Cary** and then it is assigned as the value of **city** in the following statements:

```
%let name=Cary;
%let city=&name;
```

Generally, the macro processor treats alphabetic characters, digits, and symbols (except & and %) as characters. It can also treat & and % as characters using a special treatment, which is described later. It does not make a distinction between character and numeric values as the rest of SAS does. (However, the "%EVAL Function" on page 243 and "%SYSEVALF Function" on page 263 can evaluate macro variables as integers or floating point numbers.)

Macro variable values can represent text to be generated by the macro processor or text to be used by the macro processor. Values can range in length from 0 to 65,534 characters. If you omit the value argument, the value is null (0 characters). By default, leading and trailing blanks are not stored with the value.

In addition to the %LET statement, the following list contains other features of the macro language that create macro variables:

- iterative %DO statement
- %GLOBAL statement
- %INPUT statement
- INTO clause of the SELECT statement in SQL
- %LOCAL statement
- %MACRO statement
- SYMPUT and SYMPUTX routine and SYMPUTN routine in SCL
- %WINDOW statement

The following table describes how to assign a variety of types of values to macro variables.

Table 3.2 *Types of Assignments for Macro Variable Values*

Assign	Values

Constant text	A character string. The following statements show several ways that the value **`maple`** can be assigned to macro variable STREET. In each case, the macro processor stores the five-character value **`maple`** as the value of STREET. The leading and trailing blanks are not stored.
	`%let street=maple;`
	
	`%let street=           maple;`
	`%let street=maple           ;`
	Note: Quotation marks are not required. If quotation marks are used, they become part of the value.

Digits	The appropriate digits. This example creates the macro variables NUM and TOTALSTR:
	`%let num=123;`
	`%let totalstr=100+200;`
	The macro processor does not treat **`123`** as a number or evaluate the expression **`100+200`**. Instead, the macro processor treats all the digits as characters.

Arithmetic expressions	The %EVAL function, for example,
	`%let num=%eval(100+200); / * produces 300 * /`
	use the %SYSEVALF function, for example,
	`%let num=%sysevalf(100+1.597); / * produces 101.597 * /`
	For more information, see "Macro Evaluation Functions" on page 160.

A null value	No assignment for the value argument. For example, `%let country=;`

A macro variable reference	A macro variable reference, *¯o-variable*. For example,
	`%let street=Maple;`
	`%let num=123;`
	`%let address=&num &street Avenue;`
	This example shows multiple macro references that are part of a text expression. The macro processor attempts to resolve text expressions before it makes the assignment. Thus, the macro processor stores the value of macro variable ADDRESS as **`123 Maple Avenue`**.
	You can treat ampersands and percent signs as literals by using the %NRSTR function to mask the character so that the macro processor treats it as text instead of trying to interpret it as a macro call. See "Macro Language Elements" on page 155 and Macro Quoting on page 80 for information.

<table>
<tr><th>Assign</th><th>Values</th></tr>
<tr><td>A macro invocation</td><td>

A macro call, %*macro-name*. For example,

```
%let status=%wait;
```

When the %LET statement executes, the macro processor also invokes the macro WAIT. The macro processor stores the text produced by the macro WAIT as the value of STATUS.

To prevent the macro from being invoked when the %LET statement executes, use the %NRSTR function to mask the percent sign:

```
%let status=%nrstr(%wait);
```

The macro processor stores **`%wait`** as the value of STATUS.

</td></tr>
<tr><td>Blanks and special characters</td><td>

Macro quoting function %STR or %NRSTR around the value. This action masks the blanks or special characters so that the macro processor interprets them as text. See “Macro Quoting Functions” on page 161. For example,

```
%let state=%str( North Carolina);
%let town=%str(Taylor%'s Pond);
%let store=%nrstr(Smith&Jones);
%let plotit=%str(
       proc plot;
         plot income*age;
       run;);
```

The definition of macro variable TOWN demonstrates using %STR to mask a value containing an unmatched quotation mark. “Macro Quoting Functions” on page 161 discuss macro quoting functions that require unmatched quotation marks and other symbols to be marked.

The definition of macro variable PLOTIT demonstrates using %STR to mask blanks and special characters (semicolons) in macro variable values. When a macro variable contains complete SAS statements, the statements are easier to read if you enter them on separate lines with indentions for statements within a DATA or PROC step. Using a macro quoting function retains the significant blanks in the macro variable value.

</td></tr>
<tr><td>Value from a DATA step</td><td>

The SYMPUT routine. This example puts the number of observations in a data set into a FOOTNOTE statement where AGE is greater than 20:

```
data _null_;
   set in.permdata end=final;
   if age>20 then n+1;
   if final then call symput('number',trim(left(n)));
run;
footnote "&number Observations have AGE>20";
```

During the last iteration of the DATA step, the SYMPUT routine creates a macro variable named NUMBER whose value is the value of N. (SAS also issues a numeric-to-character conversion message.) The TRIM and the LEFT functions remove the extra space characters from the DATA step variable N before its value is assigned to the macro variable NUMBER.

For a discussion of SYMPUT, including information about preventing the numeric-character message, see “CALL SYMPUT Routine” on page 224.

</td></tr>
</table>

Using Macro Variables

Macro Variable Reference

After a macro variable is created, you typically use the variable by referencing it with an ampersand preceding its name (*&variable-name*), which is called a *macro variable reference*. These references perform symbolic substitutions when they resolve to their value. You can use these references anywhere in a SAS program. To resolve a macro variable reference that occurs within a literal string, enclose the string in double quotation marks. Macro variable references that are enclosed in single quotation marks are not resolved. Compare the following statements that assign a value to macro variable DSN and use it in a TITLE statement:

```
%let dsn=Newdata;
title1 "Contents of Data Set &dsn";
title2 'Contents of Data Set &dsn';
```

In the first TITLE statement, the macro processor resolves the reference by replacing &DSN with the value of macro variable DSN. In the second TITLE statement, the value for DSN does not replace &DSN. SAS sees the following statements:

```
TITLE1 "Contents of Data Set Newdata";
TITLE2 'Contents of Data Set &dsn';
```

You can refer to a macro variable as many times as you need to in a SAS program. The value remains constant until you change it. For example, this program refers to macro variable DSN twice:

```
%let dsn=Newdata;
data temp;
   set &dsn;
   if age>=20;
run;

proc print;
   title "Subset of Data Set &dsn";
run;
```

Each time the reference &DSN appears, the macro processor replaces it with **`Newdata`**. SAS sees the following statements:

```
DATA TEMP;
      SET NEWDATA;
      IF AGE>=20;
   RUN;

   PROC PRINT;
      TITLE "Subset of Data Set NewData";
   RUN;
```

Note: If you reference a macro variable that does not exist, a warning message is printed in the SAS log. For example, if macro variable JERRY is misspelled as JERY, the following produces an unexpected result:

```
%let jerry=student;
data temp;
```

```
  x="produced by &jery";
run;
```

This code produces the following message:

```
WARNING:  Apparent symbolic reference JERY not resolved.
```

Combining Macro Variable References with Text

It is often useful to place a macro variable reference next to leading or trailing text (for example, DATA=PERSNL&YR.EMPLOYES, where &YR contains two characters for a year), or to reference adjacent variables (for example, &MONTH&YR). You can reuse the same text in several places or to reuse a program because you can change values for each use.

To reuse the same text in several places, you can write a program with macro variable references representing the common elements. You can change all the locations with a single %LET statement, as shown:

```
%let name=sales;
   data new&name;
      set save.&name;
      more SAS statements
      if units>100;
   run;
```

After macro variable resolution, SAS sees these statements:

```
DATA NEWSALES;
      SET SAVE.SALES;
      more SAS statements
      IF UNITS>100;
   RUN;
```

Notice that macro variable references do not require the concatenation operator as the DATA step does. SAS forms the resulting words automatically.

Delimiting Macro Variable Names within Text

Sometimes when you use a macro variable reference as a prefix, the reference does not resolve as you expect if you simply concatenate it. Instead, you might need to delimit the reference by adding a period to the end of it.

A period immediately following a macro variable reference acts as a delimiter. That is, a period at the end of a reference forces the macro processor to recognize the end of the reference. The period does not appear in the resulting text.

Continuing with the example above, suppose that you need another DATA step that uses the names SALES1, SALES2, and INSALES.TEMP. You might add the following step to the program:

```
/*  first attempt to add suffixes--incorrect  */
data &name1 &name2;
   set in&name.temp;
run;
```

After macro variable resolution, SAS sees these statements:

```
DATA &NAME1 &NAME2;
   SET INSALESTEMP;
RUN;
```

None of the macro variable references have resolved as you intended. The macro processor issues warning messages, and SAS issues syntax error messages. Why?

Because NAME1 and NAME2 are valid SAS names, the macro processor searches for those macro variables rather than for NAME, and the references pass into the DATA statement without resolution.

In a macro variable reference, the word scanner recognizes that a macro variable name has ended when it encounters a character that is not used in a SAS name. However, you can use a period (.) as a delimiter for a macro variable reference. For example, to cause the macro processor to recognize the end of the word NAME in this example, use a period as a delimiter between &NAME and the suffix:

```
/*  correct version  */
data &name.1 &name.2;
```

SAS now sees this statement:

```
DATA SALES1 SALES2;
```

Creating a Period to Follow Resolved Text

Sometimes you need a period to follow the text resolved by the macro processor. For example, a two-level data set name needs to include a period between the libref and data set name.

When the character following a macro variable reference is a period, use two periods. The first is the delimiter for the macro reference, and the second is part of the text.

```
set in&name..temp;
```

After macro variable resolution, SAS sees this statement:

```
SET INSALES.TEMP;
```

You can end any macro variable reference with a delimiter, but the delimiter is necessary only if the characters that follow can be part of a SAS name. For example, both of these TITLE statements are correct:

```
title "&name.--a report";
   title "&name--a report";
```

They produce the following:

```
TITLE "sales--a report";
```

Displaying Macro Variable Values

The simplest way to display macro variable values is to use the %PUT statement, which writes text to the SAS log. For example, the following statements write the following result:

```
%let a=first;
%let b=macro variable;
%put &a ***&b***;
```

Here is the result:

```
first ***macro variable***
```

You can also use a "%PUT Statement" on page 310 to view available macro variables. %PUT provides several options that enable you to view individual categories of macro variables.

The system option SYMBOLGEN displays the resolution of macro variables. For this example, assume that macro variables PROC and DSET have the values GPLOT and SASUSER.HOUSES, respectively.

```
options symbolgen;
title "%upcase(&proc) of %upcase(&dset)";
```

The SYMBOLGEN option prints to the log:

```
SYMBOLGEN:  Macro variable PROC resolves to gplot
SYMBOLGEN:  Macro variable DSET resolves to sasuser.houses
```

For more information about debugging macro programs, see "Macro Facility Error Messages and Debugging" on page 120.

Referencing Macro Variables Indirectly

Using an Expression to Generate a Reference

The macro variable references shown so far have been direct macro references that begin with one ampersand: *&name*. However, it is also useful to be able to indirectly reference macro variables that belong to a series so that the name is determined when the macro variable reference resolves. The macro facility provides indirect macro variable referencing, which enables you to use an expression (for example, CITY&N) to generate a reference to one of a series of macro variables. For example, you could use the value of macro variable N to reference a variable in the series of macro variables named CITY1 to CITY20. If N has the value 8, the reference would be to CITY8. If the value of N is 3, the reference would be to CITY3.

Although for this example the type of reference that you want is CITY&N, the following example will not produce the results that you expect, which is the value of &N appended to CITY:

```
%put &city&n;  /* incorrect */
```

This code produces a warning message saying that there is no macro variable CITY because the macro facility has tried to resolve &CITY and then &N and concatenate those values.

When you use an indirect macro variable reference, you must force the macro processor to scan the macro variable reference more than once and resolve the desired reference on the second, or later, scan. To force the macro processor to rescan a macro variable reference, you use more than one ampersand in the macro variable reference. When the macro processor encounters multiple ampersands, its basic action is to resolve two ampersands to one ampersand. For example, for you to append the value of &N to CITY and then reference the appropriate variable name, do the following:

```
%put &&city&n;  /* correct */
```

If &N contains 6, when the macro processor receives this statement, it performs the following steps:

1. resolves && to &
2. passes CITY as text
3. resolves &N into 6
4. returns to the beginning of the macro variable reference, &CITY6, starts resolving from the beginning again, and prints the value of CITY6

Generating a Series of Macro Variable References with a Single Macro Call

Using indirect macro variable references, you can generate a series of references with a single macro call by using an iterative %DO loop. The following example assumes that the macro variables CITY1 through CITY10 contain the respective values Cary, New York, Chicago, Los Angeles, Austin, Boston, Orlando, Dallas, Knoxville, and Asheville:

```
%macro listthem;
   %do n=1 %to 10; &&city&n
   %end;
%mend listthem;

%put %listthem;
```

This program writes the following to the SAS log:

```
   Cary     New York     Chicago     Los Angeles     Austin     Boston
   Orlando       Dallas       Knoxville       Asheville
```

Using More Than Two Ampersands

You can use any number of ampersands in an indirect macro variable reference, although using more than three is rare. Regardless of how many ampersands are used in this type of reference, the macro processor performs the following steps to resolve the reference.

```
%let var=city;
%let n-6;
%put &&&var&n;
```

1. It resolves the entire reference from left-to-right. If a pair of ampersands (&&) is encountered, the pair is resolved to a single ampersand, then the next part of the reference is processed. In this example, &&&VAR&N becomes &CITY6.
2. It returns to the beginning of the preliminary result and starts resolving again from left-to-right. When all ampersands have been fully processed, the resolution is complete. In this example, &CITY6 resolves to Boston, and the resolution process is finished.

Note: A macro call cannot be part of the resolution during indirect macro variable referencing.

TIP: In some cases, using indirect macro references with triple ampersands increases the efficiency of the macro processor. For more information, see "Writing Efficient and Portable Macros" on page 141.

Manipulating Macro Variable Values with Macro Functions

When you define macro variables, you can include macro functions in the expressions to manipulate the value of the variable before the value is stored. For example, you can use functions that scan other values, evaluate arithmetic and logical expressions, and remove the significance of special characters such as unmatched quotation marks.

To scan for words in macro variable values, use the %SCAN function:

```
%let address=123 maple avenue;
%let frstword=%scan(&address,1);
```

The first %LET statement assigns the string **123 maple avenue** to macro variable ADDRESS. The second %LET statement uses the %SCAN function to search the source (first argument) and retrieve the first word (second argument). Because the macro processor executes the %SCAN function before it stores the value, the value of FRSTWORD is the string **123**.

For more information about %SCAN, see "%SCAN and %QSCAN Functions" on page 250. For more information about macro functions, see "Macro Language Elements" on page 155.

Chapter 4
Macro Processing

Macro Processing

This section describes macro processing and shows the typical pattern that SAS follows to process a program containing macro elements. For most macro programming, you do not need this level of detail. It is provided to help you understand what is going on behind the scenes.

Defining and Calling Macros

Macros are compiled programs that you can call in a submitted SAS program or from a SAS command prompt. Like macro variables, you generally use macros to generate text. However, macros provide additional capabilities:

- Macros can contain programming statements that enable you to control how and when text is generated.
- Macros can accept parameters. You can write generic macros that can serve a number of uses.

To compile a macro, you must submit a macro definition. The following is the general form of a macro definition:

%MACRO *macro_name*;
<*macro_text*>

%MEND <*macro_name*>;

macro_name is a unique SAS name that identifies the macro and *macro_text* is any combination of macro statements, macro calls, text expressions, or constant text.

When you submit a macro definition, the macro processor compiles the definition and produces a member in the session catalog. The member consists of compiled macro program statements and text. The distinction between compiled items and noncompiled (text) items is important for macro execution. Examples of text items include:

- macro variable references
- nested macro calls
- macro functions, except %STR and %NRSTR
- arithmetic and logical macro expressions
- text to be written by %PUT statements
- field definitions in %WINDOW statements
- model text for SAS statements and SAS windowing environment commands

When you want to call the macro, you use the form

%macro_name.

How the Macro Processor Compiles a Macro Definition

When you submit a SAS program, the contents of the program goes to an area of memory called the input stack. The example program in the following figure contains a macro definition, a macro call, and a PROC PRINT step. This section illustrates how the macro definition in the example program is compiled and stored.

Figure 4.1 *The Macro APP*

Input Stack

```
%macro app(goal);
   %if &sysday=Friday %then
      %do;
         data thisweek;
           set lastweek;
           if totsales > &goal
             then bonus = .03;
           else bonus = 0;
      %end;
%mend app;
%app(10000)
proc print;
run;
```

Using the same process described in “SAS Programs and Macro Processing” on page 11 the word scanner begins tokenizing the program. When the word scanner detects % followed by a nonblank character in the first token, it triggers the macro processor. The macro processor examines the token and recognizes the beginning of a macro definition. The macro processor pulls tokens from the input stack and compiles until the %MEND statement terminates the macro definition (Figure 4.2 on page 35).

During macro compilation, the macro processor does the following:

- creates an entry in the session catalog

- compiles and stores all macro program statements for that macro as macro instructions
- stores all noncompiled items in the macro as text

  Note: Text items are underlined in the illustrations in this section.

If the macro processor detects a syntax error while compiling the macro, it checks the syntax in the rest of the macro and issues messages for any additional errors that it finds. However, the macro processor does not store the macro for execution. A macro that the macro processor compiles but does not store is called a *dummy macro*.

Figure 4.2 *Macro APP in the Input Stack*

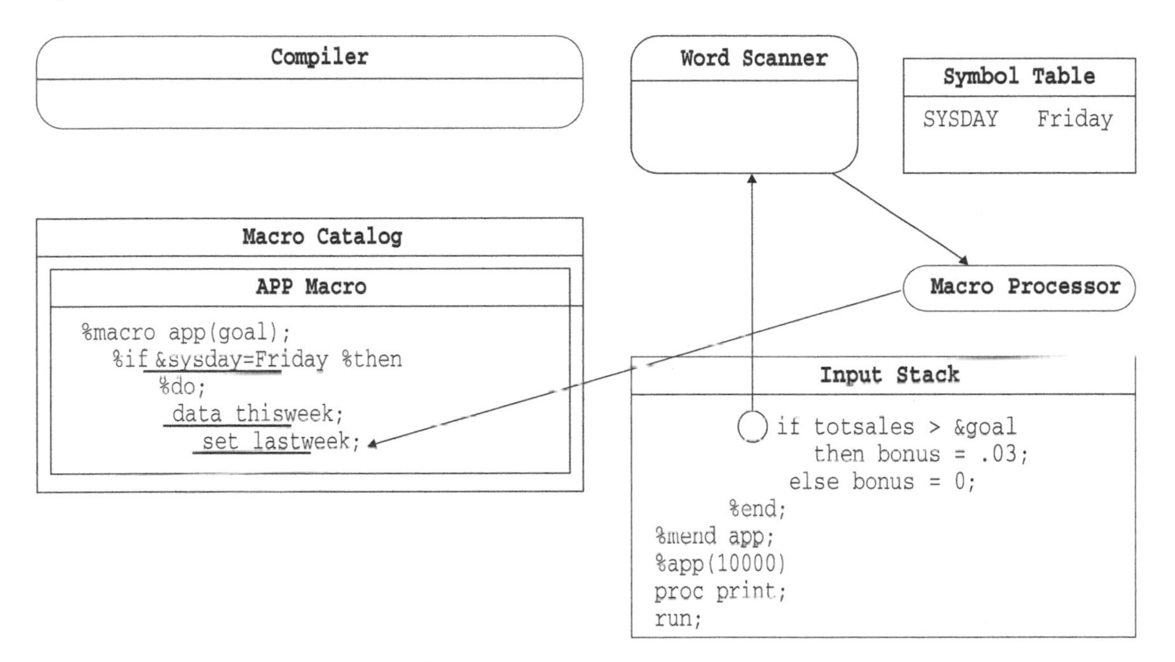

In this example, the macro definition is compiled and stored successfully. (See the following figure.) For the sake of illustration, the compiled APP macro looks like the original macro definition that was in the input stack. The entry would actually contain compiled macro instructions with constant text. The constant text in this example is underlined.

Figure 4.3 *The Compiled Macro APP*

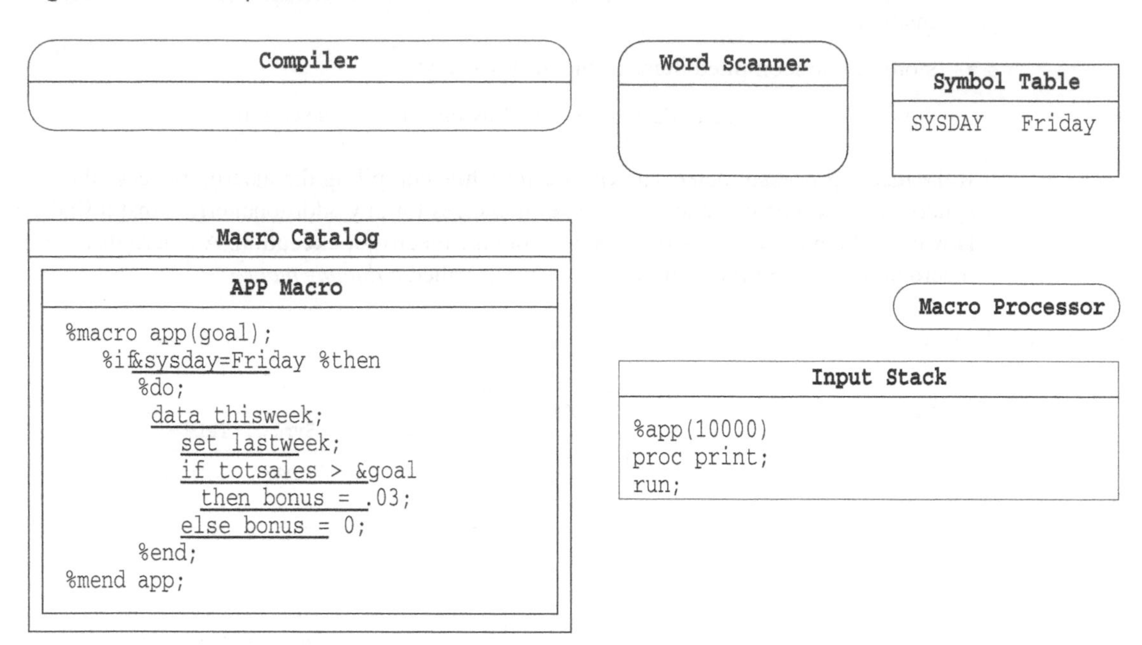

How the Macro Processor Executes a Compiled Macro

Macro execution begins with the macro processor opening the SASMACR catalog to read the appropriate macro entry. As the macro processor executes the compiled instructions in the macro entry, it performs a series of simple repetitive actions. During macro execution, the macro processor does the following:

- executes compiled macro program instructions
- places noncompiled constant text on the input stack
- waits for the word scanner to process the generated text
- resumes executing compiled macro program instructions

To continue the example from the previous section, the following figure shows the lines remaining in the input stack after the macro processor compiles the macro definition APP.

Figure 4.4 *The Macro Call in the Input Stack*

Input Stack
%app(10000) proc print; run;

The word scanner examines the input stack and detects % followed by a nonblank character in the first token. It triggers the macro processor to examine the token.

Figure 4.5 *Macro Call Entering Word Queue*

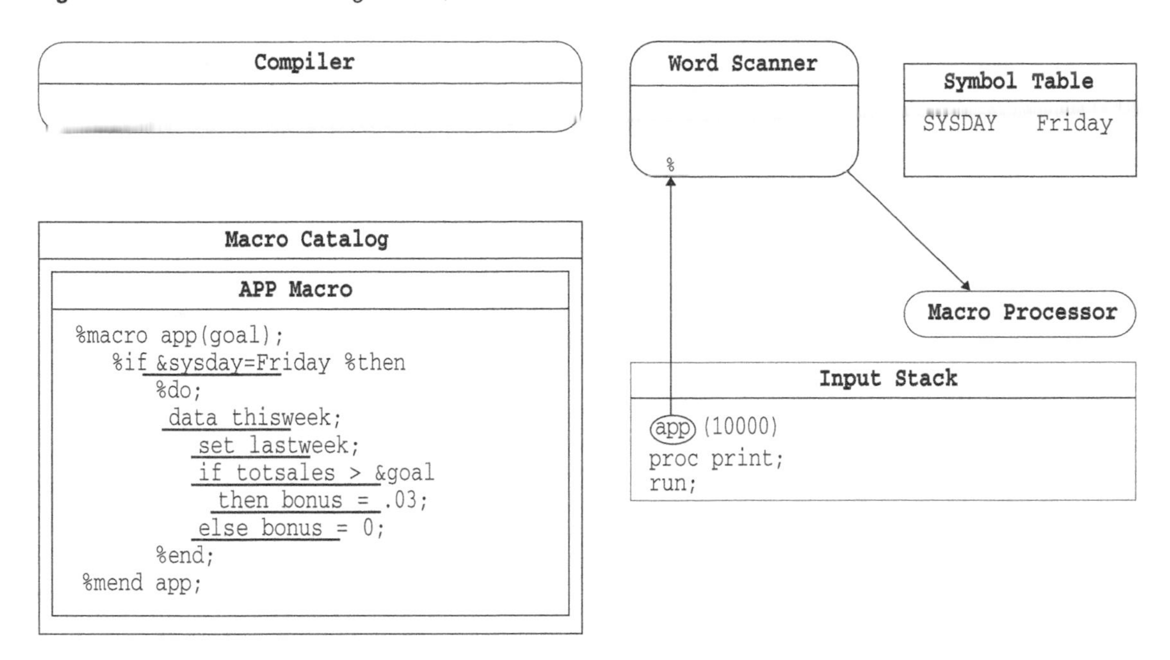

The macro processor recognizes a macro call and begins to execute macro APP, as follows:

1. The macro processor creates a local symbol table for the macro. The macro processor examines the previously compiled definition of the macro. If there are any parameters, variable declarations, or computed GOTO statements in the macro definition, the macro processor adds entries for the parameters and variables to the newly created local symbol table.

2. The macro processor further examines the previously compiled macro definition for parameters to the macro. If no parameters were defined in the macro definition, the macro processor begins to execute the compiled instructions of the macro. If any parameters were contained in the definition, the macro processor removes tokens from the input stack to obtain values for positional parameters and non-default values for keyword parameters. The values for parameters found in the input stack are placed in the appropriate entry in the local symbol table.

   Note: Before executing any compiled instructions, the macro processor removes only enough tokens from the input stack to ensure that any tokens that are supplied by the user and pertain to the macro call have been removed.

3. The macro processor encounters the compiled %IF instruction and recognizes that the next item will be text containing a condition.

4. The macro processor places the text **&sysday=Friday** on the input stack ahead of the remaining text in the program. (See the following figure). The macro processor waits for the word scanner to tokenize the generated text.

***Figure 4.6** Text for %IF Condition on Input Stack*

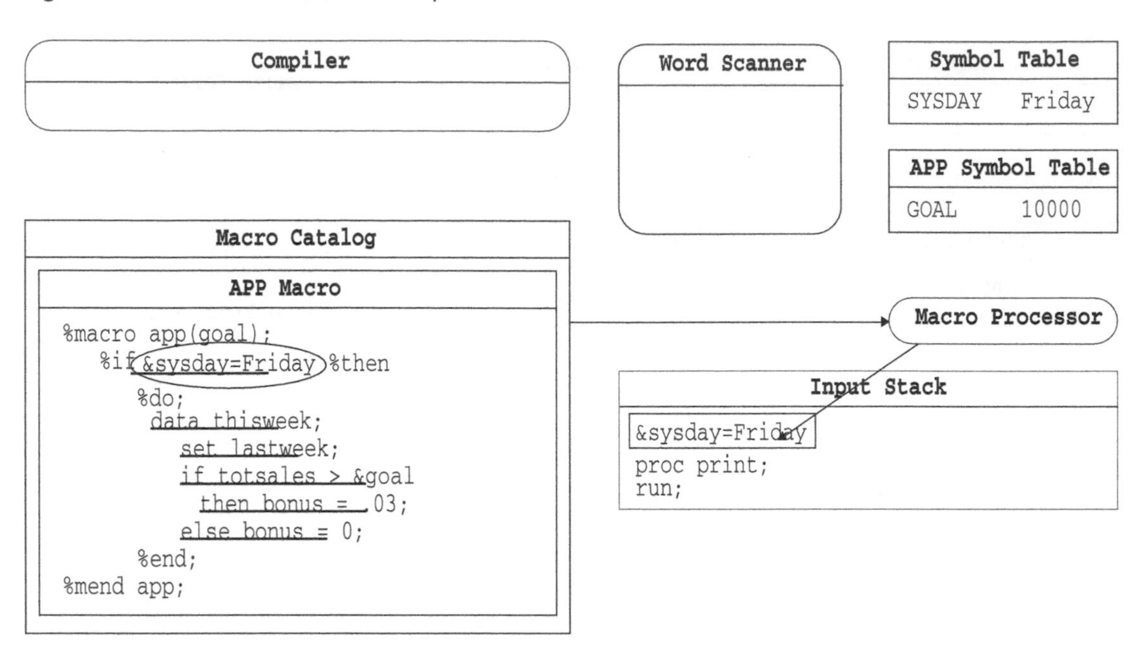

1. The word scanner starts tokenizing the generated text, recognizes an ampersand followed by nonblank character in the first token, and triggers the macro processor.
2. The macro processor examines the token and finds a possible macro variable reference, &SYSDAY. The macro processor first searches the local APP symbol table for a matching entry and then the global symbol table. When the macro processor finds the entry in the global symbol table, it replaces macro variable in the input stack with the value **`Friday`**. (See the following figure.)
3. The macro processor stops and waits for the word scanner to tokenize the generated text.

***Figure 4.7** Input Stack after Macro Variable Reference Is Resolved*

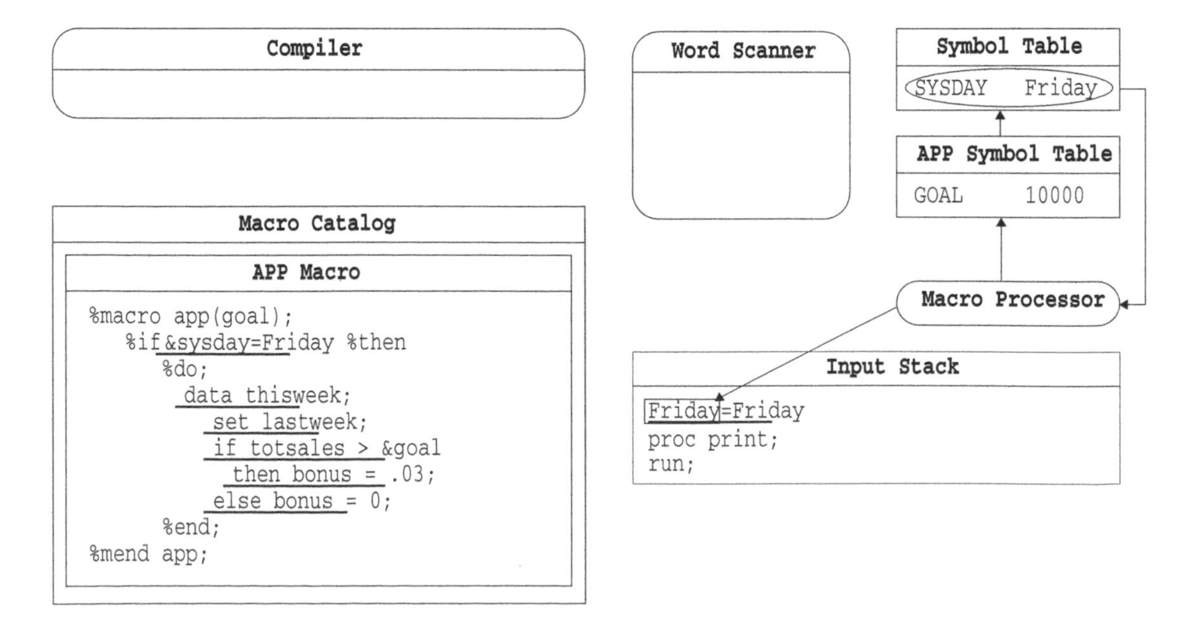

1. The word scanner then read **Friday=Friday** from the input stack.
2. The macro processor evaluates the expression **Friday=Friday** and, because the expression is true, proceeds to the %THEN and %DO instructions.

Figure 4.8 *Macro Processor Receives the Condition*

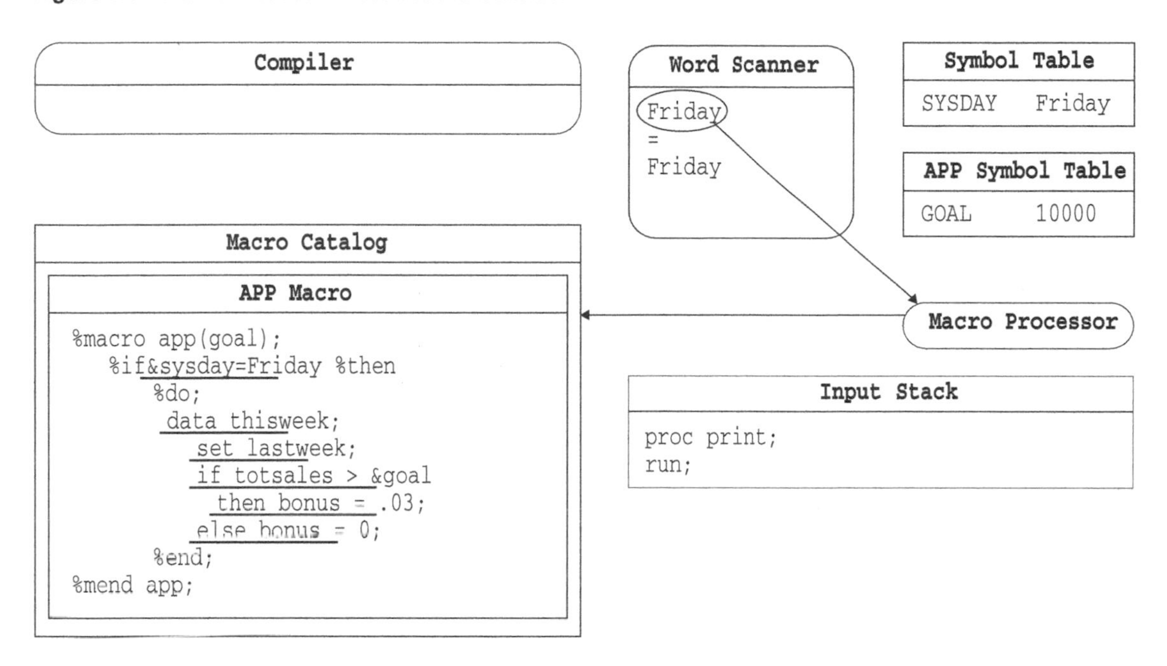

1. The macro processor executes the compiled %DO instructions and recognizes that the next item is text.
2. The macro processor places the text on top of the input stack and waits for the word scanner to begin tokenization.
3. The word scanner reads the generated text from the input stack, and tokenizes it.
4. The word scanner recognizes the beginning of a DATA step, and triggers the compiler to begin accepting tokens. The word scanner transfers tokens to the compiler from the top of the stack.

Figure 4.9 Generated Text on Top of Input Stack

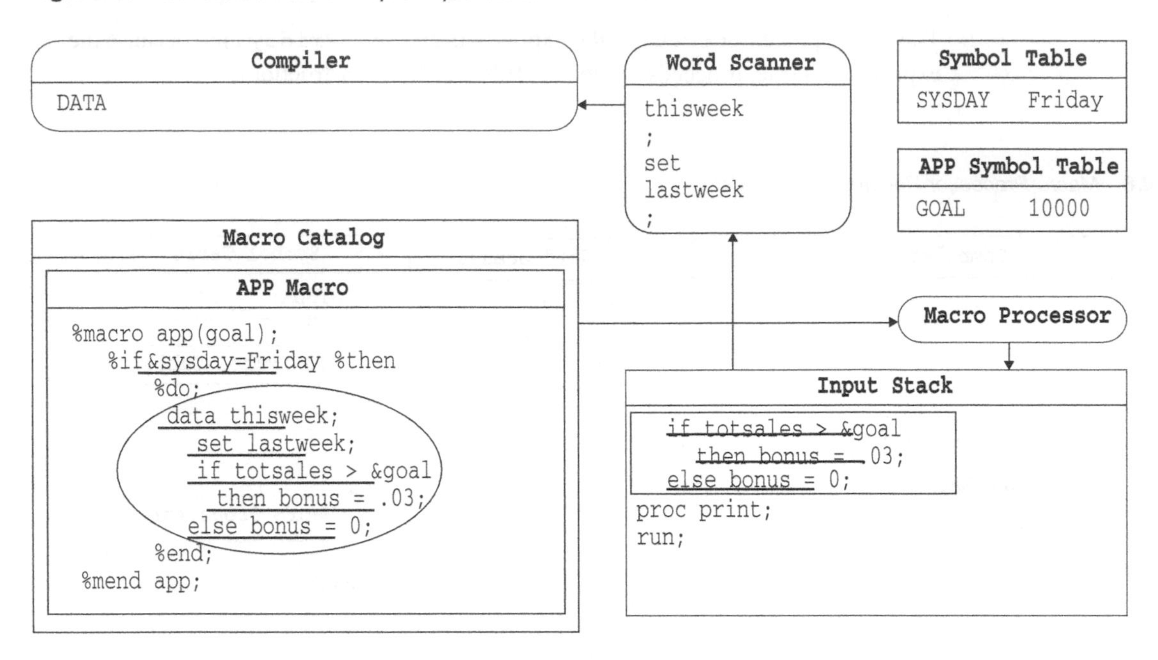

1. When the word scanner detects & followed by a nonblank character (the macro variable reference &GOAL), it triggers the macro processor.
2. The macro processor looks in the local APP symbol table and resolves the macro variable reference &GOAL to `10000`. The macro processor places the value on top of the input stack, ahead of the remaining text in the program.

Figure 4.10 The Word Scanner Reads Generated Text

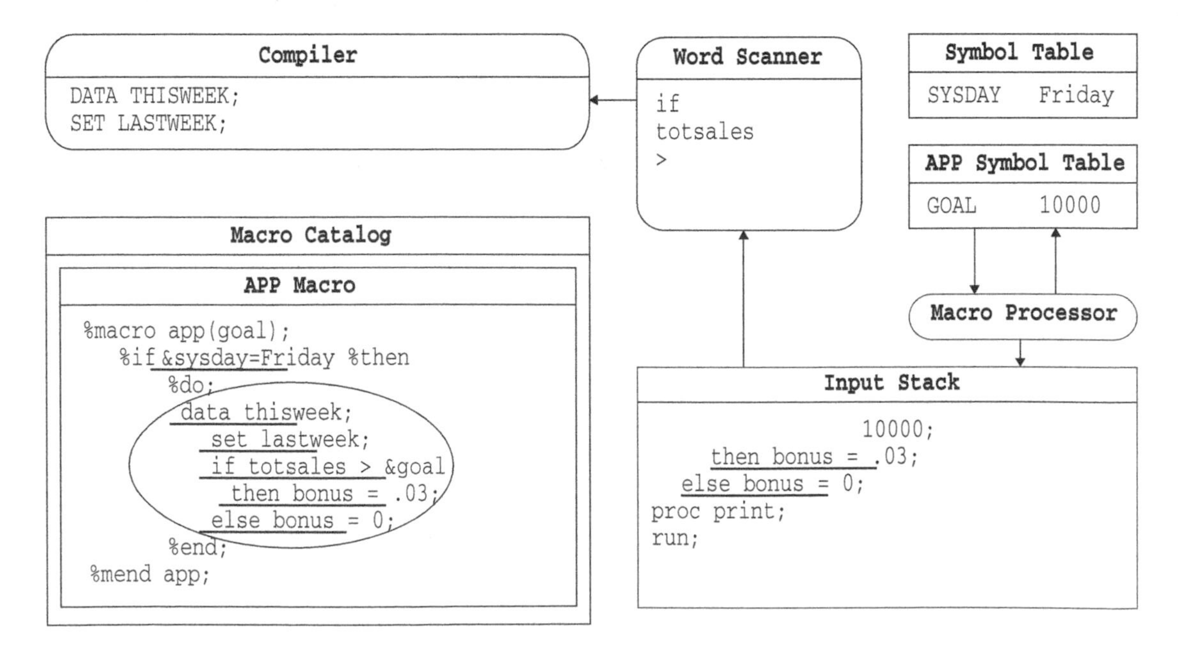

1. The word scanner resumes tokenization. When it has completed tokenizing the generated text, it triggers the macro processor.

2. The macro processor resumes processing the compiled macro instructions. It recognizes the end of the %DO group at the %END instruction and proceeds to %MEND.
3. the macro processor executes the %MEND instruction, removes the local symbol table APP, and macro APP ceases execution.
4. The macro processor triggers the word scanner to resume tokenization.
5. The word scanner reads the first token in the input stack (PROC), recognizes the beginning of a step boundary, and triggers the DATA step compiler.
6. The compiled DATA step is executed, and the DATA step compiler is cleared.
7. The word scanner signals the PRINT procedure (a separate executable not illustrated), which pulls the remaining tokens.

Figure 4.11 *The Remaining Statements Are Compiled and Executed*

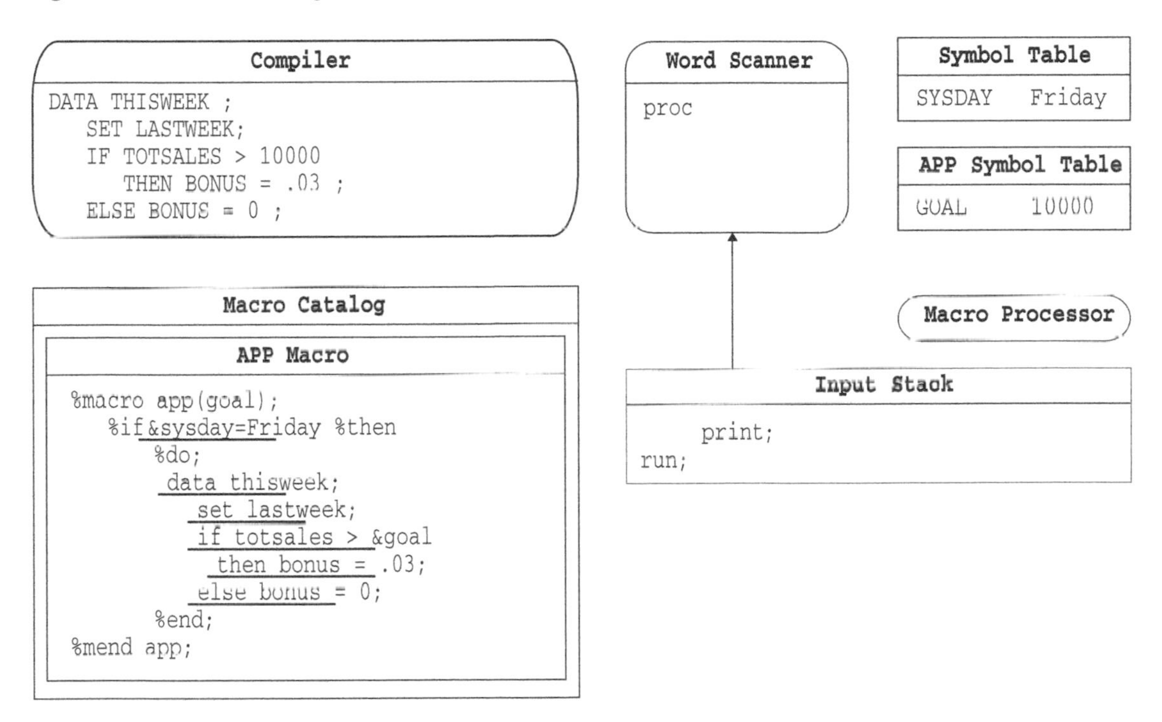

Summary of Macro Processing

The previous sections illustrate the relationship between macro compilation and execution and DATA step compilation and execution. The relationship contains a pattern of simple repetitive actions. These actions begin when text is submitted to the input stack and the word scanner begins tokenization. At times the word scanner waits for the macro processor to perform an activity, such as searching the symbol tables or compiling a macro definition. If the macro processor generates text during its activity, then it pauses while the word scanner tokenizes the text and sends the tokens to the appropriate target. These tokens might trigger other actions in parts of SAS, such as the DATA step compiler, the command processor, or a SAS procedure. If any of these actions occur, the macro processor waits for these actions to be completed before resuming its activity.

When the macro processor stops, the word scanner resumes tokenization. This process continues until the entire program has been processed.

Chapter 5
Scopes of Macro Variables

Scopes of Macro Variables

Every macro variable has a *scope*. A macro variable's scope determines how it is assigned values and how the macro processor resolves references to it.

Two types of scopes exist for macro variables: *global* and *local*. Global macro variables exist for the duration of the SAS session and can be referenced anywhere (except CARDS and DATALINES) in the program—either inside or outside a macro. Local macro variables exist only during the execution of the macro in which the variables are created and have no meaning outside the defining macro.

Scopes can be nested, like boxes within boxes. For example, suppose you have a macro A that creates the macro variable LOC1 and a macro B that creates the macro variable LOC2. If the macro B is nested (executed) within the macro A, LOC1 is local to both A and B. However, LOC2 is local only to B.

Macro variables are stored in *symbol tables*, which list the macro variable name and its value. There is a global symbol table, which stores all global macro variables. Local macro variables are stored in a local symbol table that is created at the beginning of the execution of a macro.

You can use the %SYMEXIST function to indicate whether a macro variable exists. See "%SYMEXIST Function" on page 260 for more detailed information.

Global Macro Variables

The following figure illustrates the global symbol table during execution of the following program:

```
%let county=Clark;

%macro concat;
   data _null_;
      length longname $20;
      longname="&county"||" County";
      put longname;
   run;
%mend concat;

%concat
```

Calling the macro CONCAT produces the following statements:

```
data _null_;
      length longname $20;
      longname="Clark"||" County";
      put longname;
run;
```

The PUT statement writes the following to the SAS log:

```
Clark County
```

Figure 5.1 Global Symbol Table

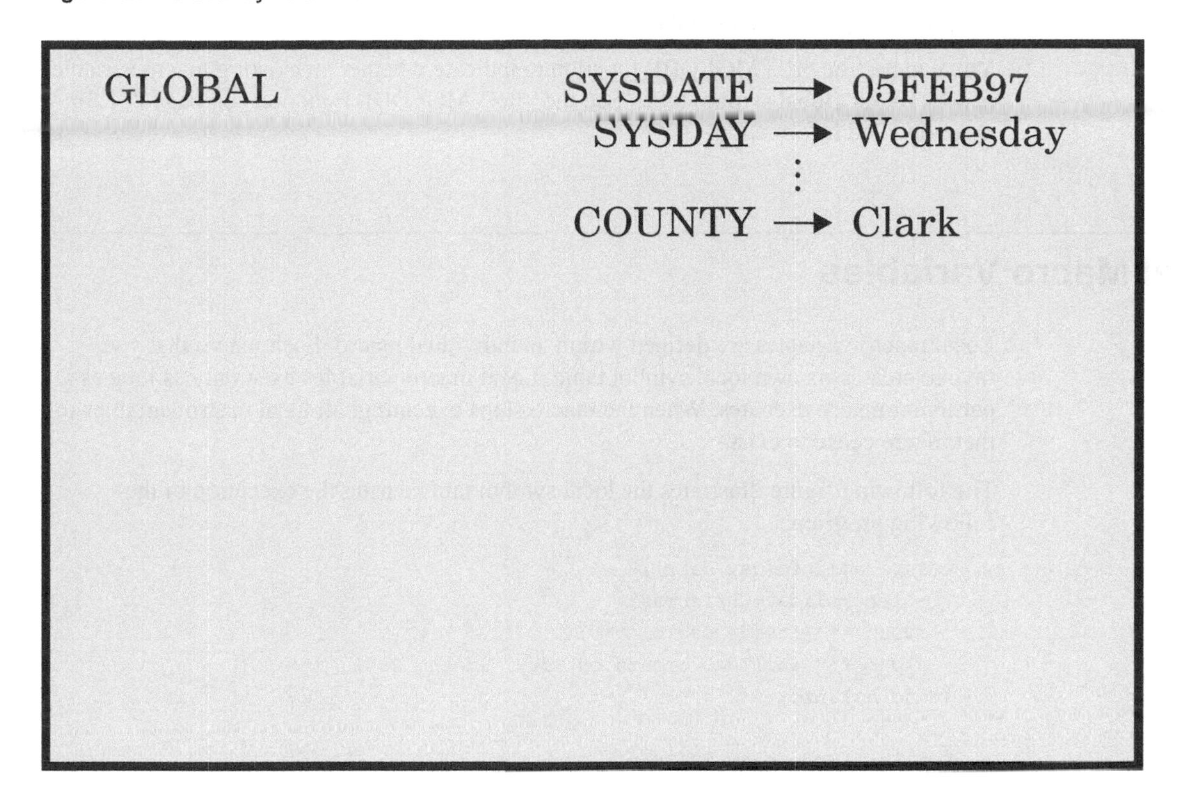

Global macro variables include the following:

- all automatic macro variables except SYSPBUFF. See "Automatic Macro Variables" on page 164 for more information about SYSPBUFF and other automatic macro variables.
- macro variables created outside of any macro.
- macro variables created in %GLOBAL statements. See "Creating Global Macro Variables" on page 61 for more information about the %GLOBAL statement.
- most macro variables created by the CALL SYMPUT routine. See "Special Cases of Scope with the CALL SYMPUT Routine" on page 63 for more information about the CALL SYMPUT routine.

You can create global macro variables any time during a SAS session or job. Except for some automatic macro variables, you can change the values of global macro variables any time during a SAS session or job.

In most cases, once you define a global macro variable, its value is available to you anywhere in the SAS session or job and can be changed anywhere. So, a macro variable referenced inside a macro definition is global if a global macro variable already exists by the same name (assuming that the variable is not specifically defined as local with the %LOCAL statement or in a parameter list). The new macro variable definition simply updates the existing global one. The following are exceptions that prevent you from referencing the value of a global macro variable:

- when a macro variable exists both in the global symbol table and in the local symbol table, you cannot reference the global value from within the macro that contains the local macro variable. In this case, the macro processor finds the local value first and uses it instead of the global value.
- if you create a macro variable in the DATA step with the SYMPUT routine, you cannot reference the value with an ampersand until the program reaches a step

boundary. See “Macro Processing” on page 33 for more information about macro processing and step boundaries.

You can use the %SYMGLOBL function to indicate whether an existing macro variable resides in the global symbol table. See the “%SYMGLOBL Function” on page 261 for more detailed information.

Local Macro Variables

Local macro variables are defined within an individual macro. Each macro that you invoke creates its own local symbol table. Local macro variables exist only as long as a particular macro executes. When the macro stops executing, all local macro variables for that macro cease to exist.

The following figure illustrates the local symbol table during the execution of the following program.

```
%macro holinfo(day,date);
   %let holiday=Christmas;
   %put *** Inside macro: ***;
   %put *** &holiday occurs on &day, &date, 2002. ***;
%mend holinfo;

%holinfo(Wednesday,12/25)

%put *** Outside macro: ***;
%put *** &holiday occurs on &day, &date, 2002. ***;
```

The %PUT statements write the following to the SAS log:

```
*** Inside macro: ***
*** Christmas occurs on Wednesday, 12/25, 2002. ***

*** Outside macro: ***
WARNING: Apparent symbolic reference HOLIDAY not resolved.
WARNING: Apparent symbolic reference DAY not resolved.
WARNING: Apparent symbolic reference DATE not resolved.
*** &holiday occurs on &day, &date, 2002. ***
```

As you can see from the log, the local macro variables DAY, DATE, and HOLIDAY resolve inside the macro, but outside the macro that they do not exist and therefore do not resolve.

Figure 5.2 Local Symbol Table

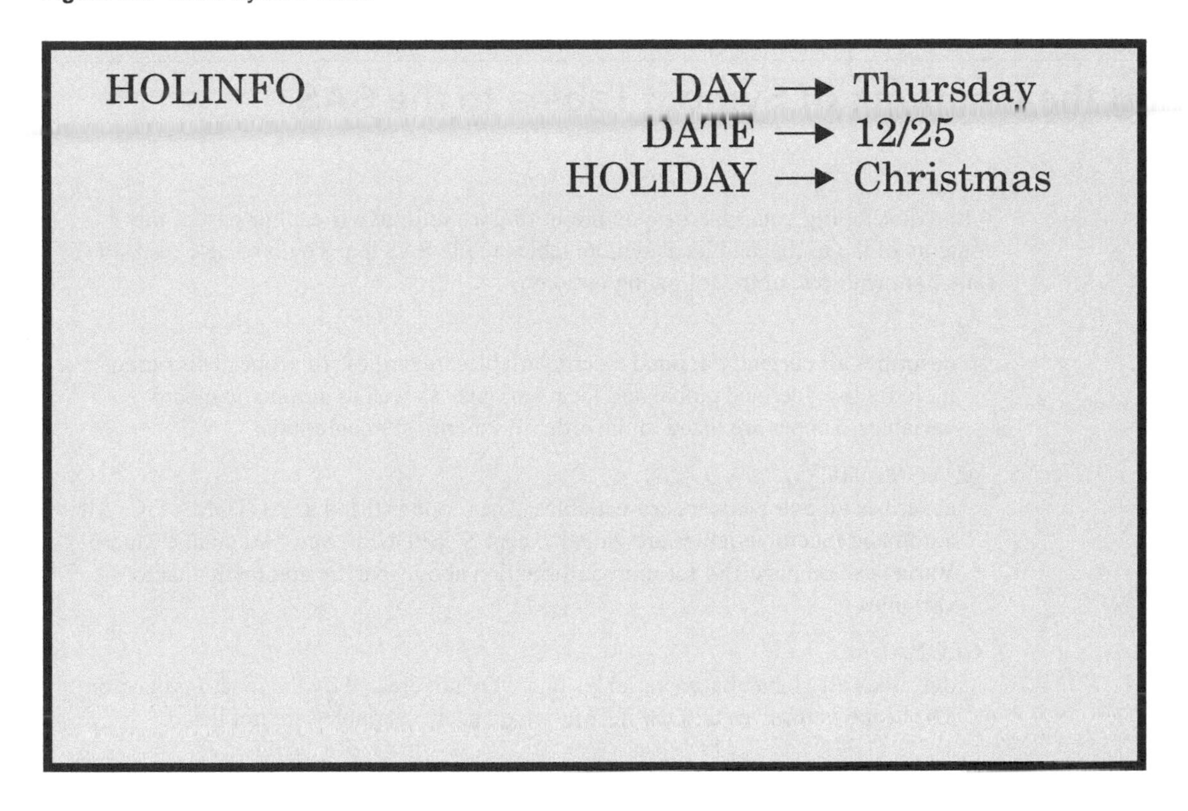

A macro's local symbol table is empty until the macro creates at least one macro variable. A local symbol table can be created by any of the following:

- the presence of one or more macro parameters
- a %LOCAL statement
- macro statements that define macro variables, such as %LET and the iterative %DO statement (if the variable does not already exist globally or a %GLOBAL statement is not used)

Note: Macro parameters are always local to the macro that defines them. You cannot make macro parameters global. (Although, you can assign the value of the parameter to a global variable. See "Creating Global Variables Based on the Value of Local Variables" on page 62.)

When you invoke one macro inside another, you create nested scopes. Because you can have any number of levels of nested macros, your programs can contain any number of levels of nested scopes.

You can use the %SYMLOCAL function to indicate whether an existing macro variable resides in an enclosing local symbol table. See the "%SYMLOCAL Function" on page 262 for more detailed information.

Writing the Contents of Symbol Tables to the SAS Log

While developing your macros, you might find it useful to write all or part of the contents of the global and local symbol tables to the SAS log. To do so, use the %PUT statement with one of the following options:

ALL
: describes all currently defined macro variables, regardless of scope. This output includes user-defined global and local variables as well as automatic macro variables. Scopes are listed in the order of innermost to outermost.

AUTOMATIC
: describes all automatic macro variables. The scope is listed as AUTOMATIC. All automatic macro variables are global except SYSPBUFF. See “Automatic Macro Variables” on page 164 for more information about specific automatic macro variables.

GLOBAL
: describes all global macro variables that were not created by the macro processor. The scope is listed as GLOBAL. Automatic macro variables are not listed.

LOCAL
: describes user-defined local macro variables defined within the currently executing macro. The scope is listed as the name of the macro in which the macro variable is defined.

USER
: describes all user-defined macro variables, regardless of scope. The scope is either GLOBAL, for global macro variables, or the name of the macro in which the macro variable is defined.

For example, consider the following program:

```
%let origin=North America;

%macro dogs(type=);
   data _null_;
      set all_dogs;
      where dogtype="&type" and dogorig="&origin";
      put breed " is for &type.";
   run;

   %put _user_;
%mend dogs;

%dogs(type=work)
```

The %PUT statement preceding the %MEND statement writes to the SAS log the scopes, names, and values of all user-generated macro variables:

```
DOGS    TYPE    work
GLOBAL    ORIGIN    North America
```

Because TYPE is a macro parameter, TYPE is local to the macro DOGS, with value **work**. Because ORIGIN is defined in open code, it is global.

How Macro Variables Are Assigned and Resolved

Before the macro processor creates a variable, assigns a value to a variable, or resolves a variable, it searches the symbol tables to determine whether the variable already exists. The search begins with the most local scope and, if necessary, moves outward to the global scope. The request to assign or resolve a variable comes from a macro variable reference in open code (outside a macro) or within a macro.

The following figure illustrates the search order the macro processor uses when it receives a macro variable reference that requests a variable be created or assigned. The figure below illustrates the process for resolving macro variable references. Both these figures represent the most basic type of search and do not apply in special cases, such as when a %LOCAL statement is used or the variable is created by CALL SYMPUT.

Figure 5.3 *Search Order When Assigning or Creating Macro Variables*

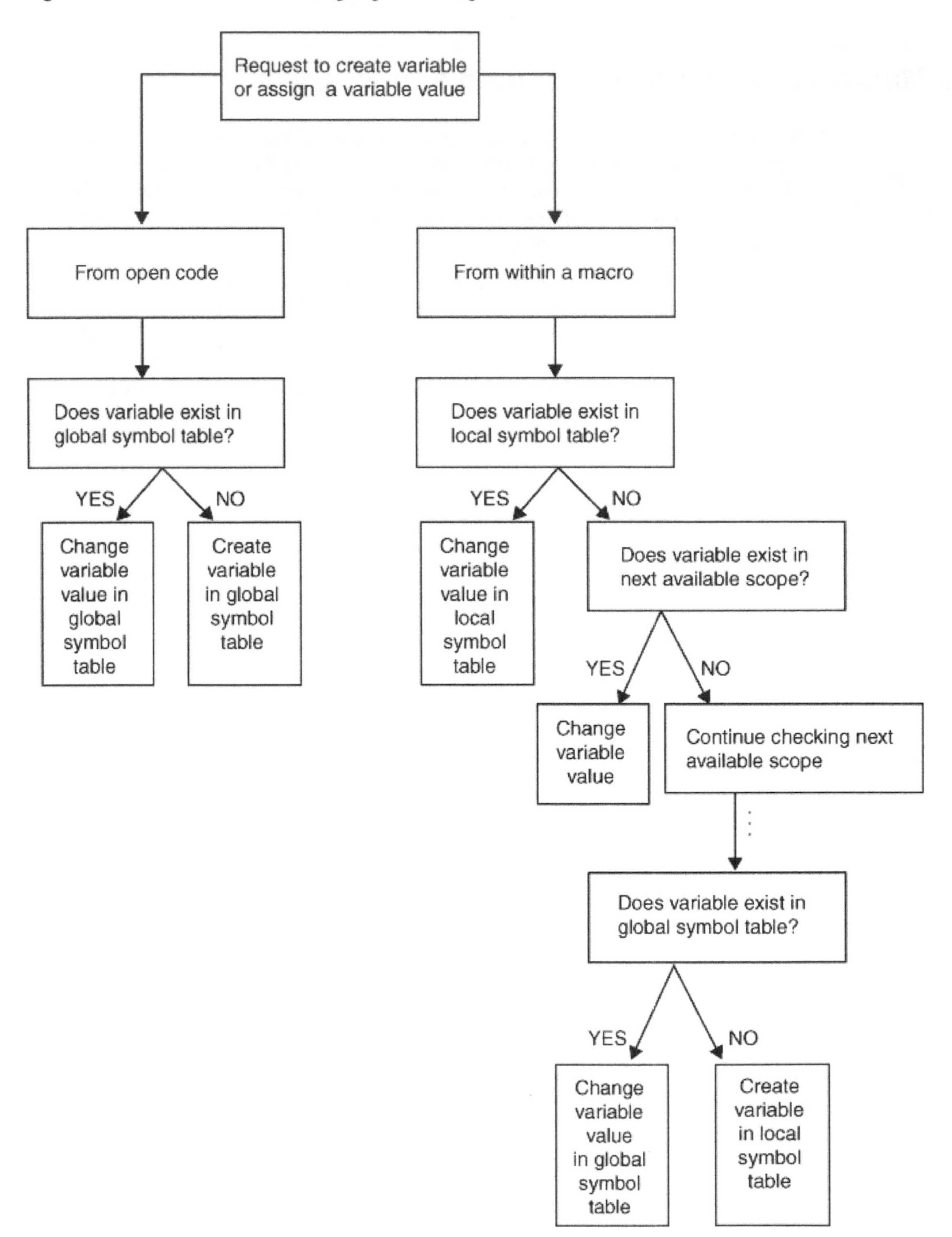

Figure 5.4 *Search Order When Resolving Macro Variable References*

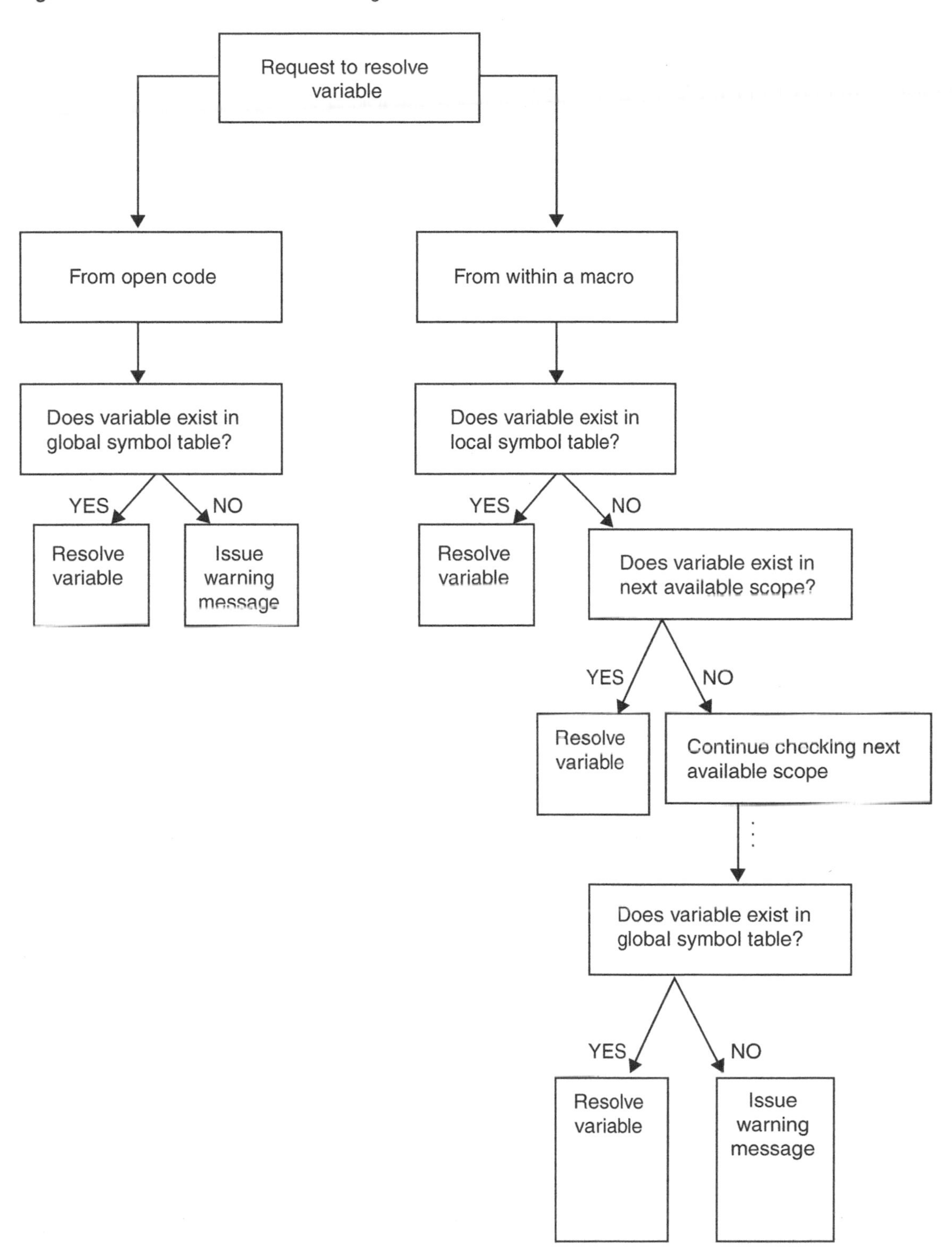

Examples of Macro Variable Scopes

Changing the Values of Existing Macro Variables

When the macro processor executes a macro program statement that can create a macro variable (such as a %LET statement), the macro processor attempts to change the value of an existing macro variable rather than create a new macro variable. The %GLOBAL and %LOCAL statements are exceptions.

To illustrate, consider the following %LET statements. Both statements assign values to the macro variable NEW:

```
%let new=inventry;
%macro name1;
   %let new=report;
%mend name1;
```

Suppose you submit the following statements:

```
%name1

data &new;

data report;
```

Because NEW exists as a global variable, the macro processor changes the value of the variable rather than creating a new one. The macro NAME1's local symbol table remains empty.

The following figure illustrates the contents of the global and local symbol tables before, during, and after NAME1's execution.

Figure 5.5 Snapshots of Symbol Tables

Before NAME1 executes

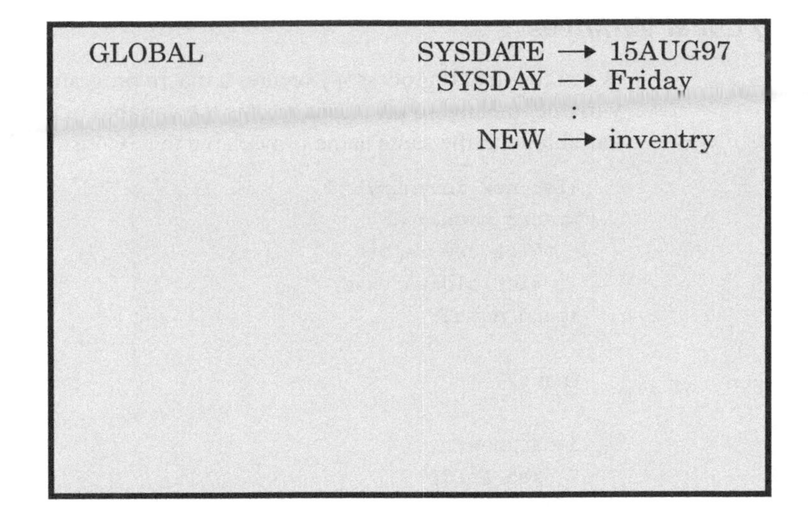

While NAME1 executes

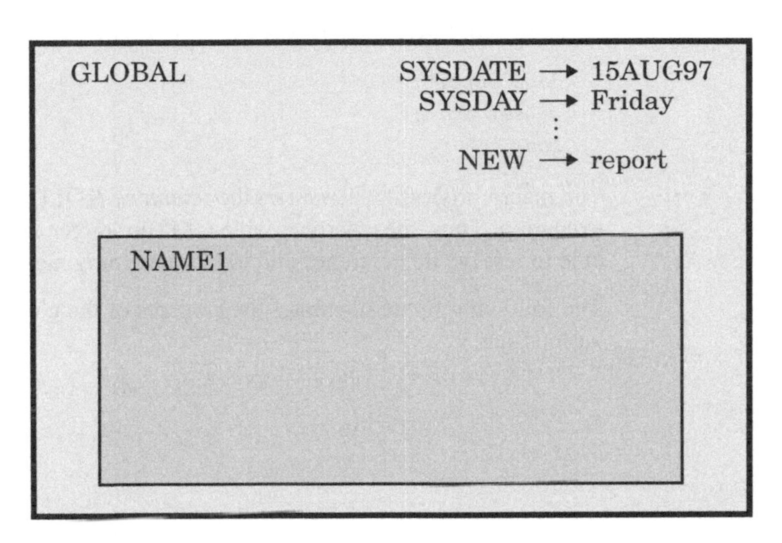

After NAME1 executes

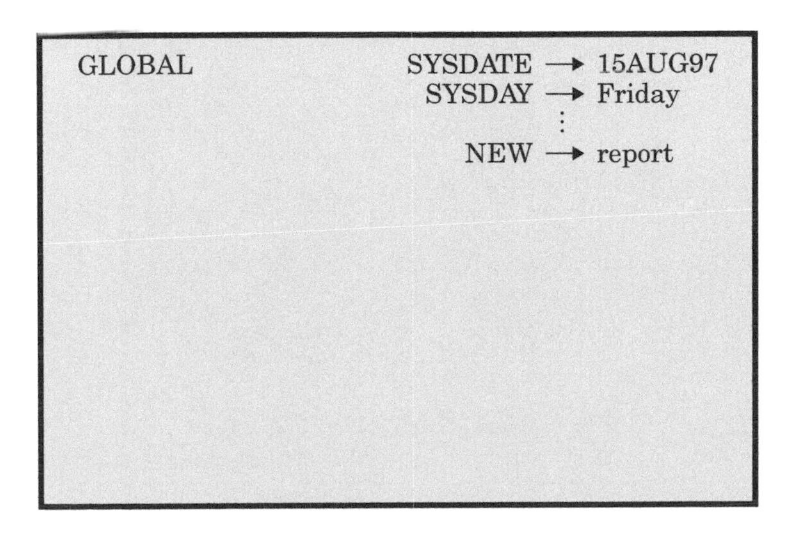

Creating Local Variables

When the macro processor executes a macro program statement that can create a macro variable, the macro processor creates the variable in the local symbol table if no macro variable with the same name is available to it. Consider the following example:

```
%let new=inventry;
%macro name2;
   %let new=report;
   %let old=warehse;
%mend name2;

%name2

data &new;
   set &old;
run;
```

After NAME2 executes, the SAS compiler sees the following statements:

```
data report;
   set &old;
run;
```

The macro processor encounters the reference &OLD after macro NAME2 has finished executing. Thus, the macro variable OLD no longer exists. The macro processor is not able to resolve the reference and issues a warning message.

The following figure illustrates the contents of the global and local symbol tables at various stages.

Figure 5.6 Symbol Tables at Various Stages

Before NAME2 executes

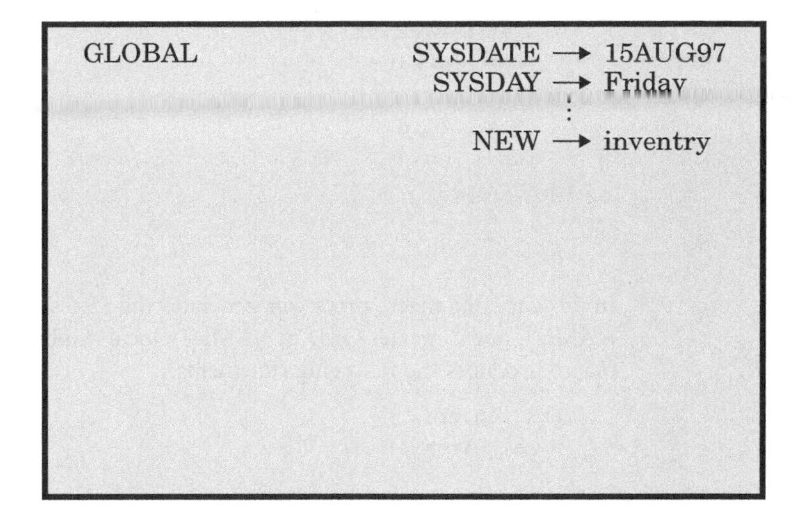

While NAME2 executes

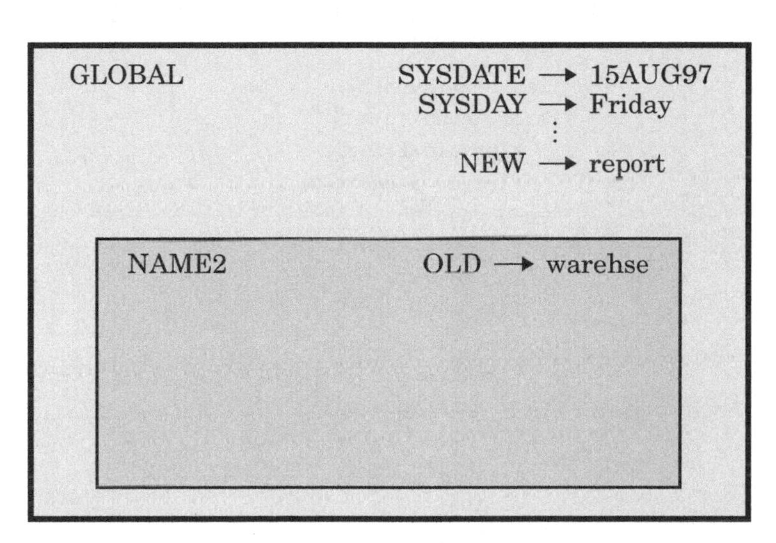

After NAME2 executes

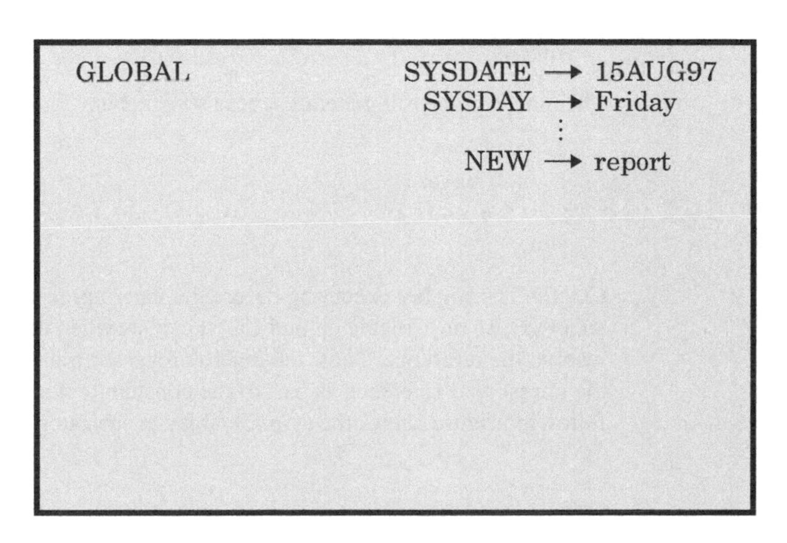

But suppose you place the SAS statements inside the macro NAME2, as in the following program:

```
%let new=inventry;
%macro name2;
   %let new=report;
   %let old=warehse;
   data &new;
      set &old;
   run;
%mend name2;

%name2
```

In this case, the macro processor generates the SET statement during the execution of NAME2, and it locates OLD in NAME2's local symbol table. Therefore, executing the macro produces the following statements:

```
data report;
   set warehse;
run;
```

The same rule applies regardless of how many levels of nesting exist. Consider the following example:

```
%let new=inventry;
%macro conditn;
   %let old=sales;
   %let cond=cases>0;
%mend conditn;

%macro name3;
   %let new=report;
   %let old=warehse;
   %conditn
      data &new;
         set &old;
         if &cond;
      run;
%mend name3;

%name3
```

The macro processor generates these statements:

```
data report;
   set sales;
   if &cond;
run;
```

CONDITN finishes executing before the macro processor reaches the reference &COND, so no variable named COND exists when the macro processor attempts to resolve the reference. Thus, the macro processor issues a warning message and generates the unresolved reference as part of the constant text and issues a warning message. The following figure shows the symbol tables at each step.

Figure 5.7 Symbol Tables Showing Two Levels of Nesting

Early execution of NAME3, before CONDITN executes

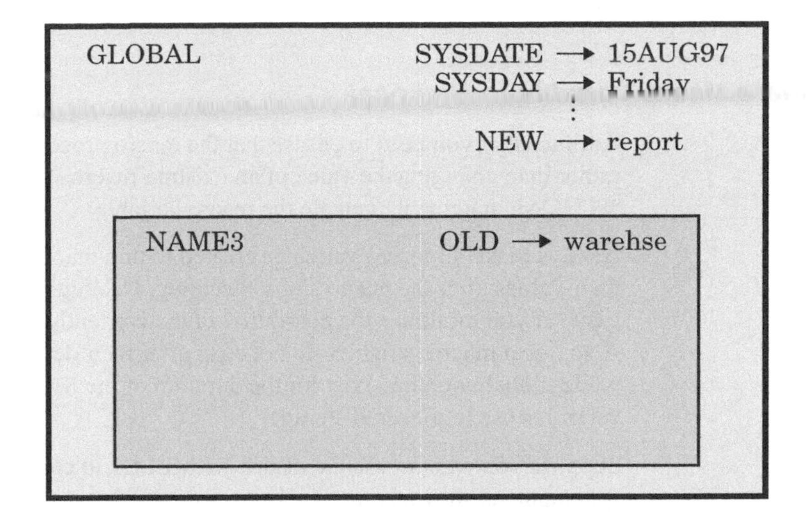

While NAME3 and CONDITN execute

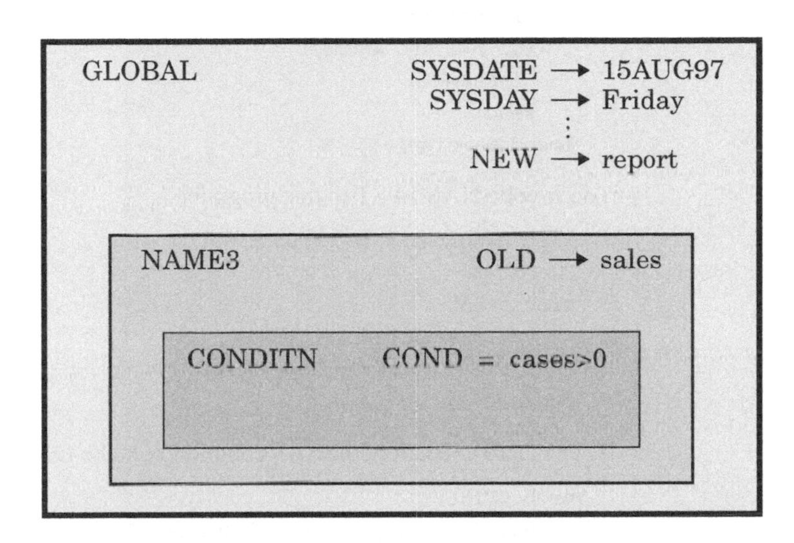

Late execution of NAME3, after CONDITN executes

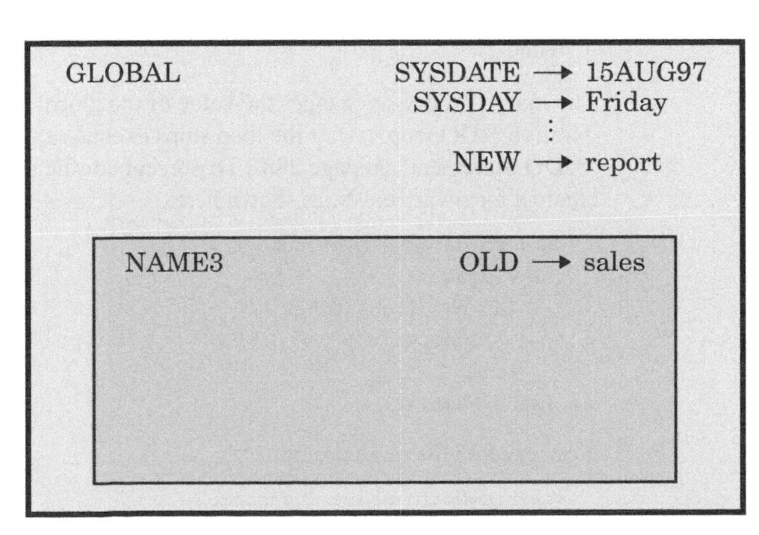

Notice that the placement of a macro invocation is what creates a nested scope, not the placement of the macro definition. For example, invoking CONDITN from within

NAME3 creates the nested scope. It is not necessary to define CONDITN within NAME3.

Forcing a Macro Variable to Be Local

At times that you need to ensure that the macro processor creates a local macro variable rather than changing the value of an existing macro variable. In this case, use the %LOCAL statement to create the macro variable.

Always make all macro variables created within macros local when you do not need their values after the macro stops executing. Debugging the large macro programs is easier if you minimize the possibility of inadvertently changing a macro variable's value. Also, local macro variables do not exist after their defining macro finishes executing, while global variables exist for the duration of the SAS session. Therefore, local variables use less overall storage.

Suppose you want to use the macro NAMELST to create a list of names for a VAR statement, as shown here:

```
%macro namelst(name,number);
   %do n=1 %to &number;
      &name&n
   %end;
%mend namelst;
```

You invoke NAMELST in this program:

```
%let n=North State Industries;

proc print;
   var %namelst(dept,5);
   title "Quarterly Report for &n";
run;
```

After macro execution, the SAS compiler sees the following statements:

```
proc print;
   var dept1 dept2 dept3 dept4 dept5;
   title "Quarterly Report for 6";
run;
```

The macro processor changes the value of the global variable N each time it executes the iterative %DO loop. (After the loop stops executing, the value of N is 6, as described in “%DO Statement” on page 288.) To prevent conflicts, use a %LOCAL statement to create a local variable N, as shown here:

```
%macro namels2(name,number);
   %local n;
   %do n=1 %to &number;
      &name&n
   %end;
%mend namels2;
```

Now execute the same program:

```
%let n=North State Industries;

proc print;
   var %namels2(dept,5);
   title "Quarterly Report for &n";
run;
```

The macro processor generates the following statements:

```
proc print;
   var dept1 dept2 dept3 dept4 dept5;
   title "Quarterly Report for North State Industries";
run;
```

The following figure shows the symbol tables before NAMELS2 executes, while NAMELS2 is executing, and when the macro processor encounters the reference &N in the TITLE statement.

Figure 5.8 *Symbol Tables for Global and Local Variables with the Same Name*

Before NAMELS2 executes

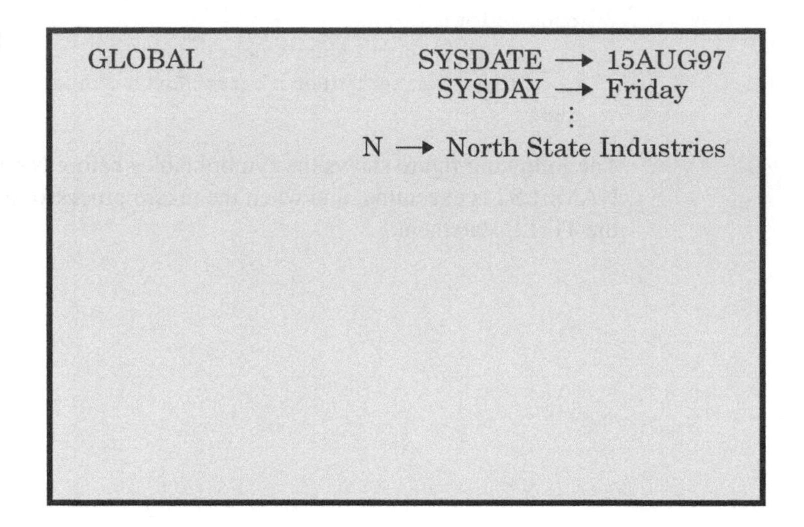

While NAMELS2 executes (at end of last iteration of %DO loop)

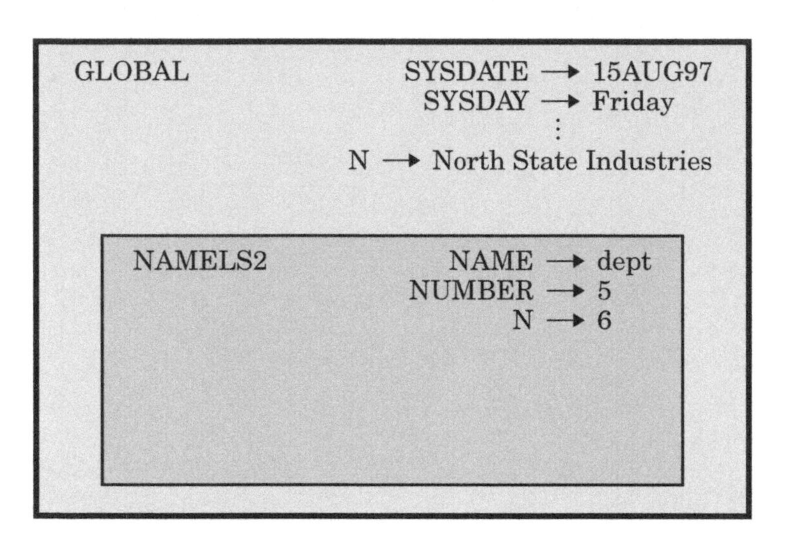

After NAMELS2 executes

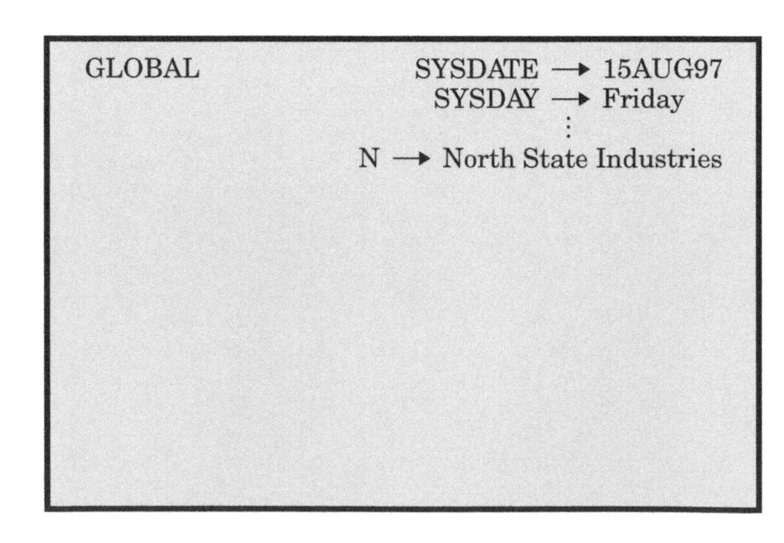

Creating Global Macro Variables

The %GLOBAL statement creates a global macro variable if a variable with the same name does not already exist there, regardless of what scope is current.

For example, in the following program, the macro CONDITN contains a %GLOBAL statement that creates the macro variable COND as a global variable:

```
%macro conditn;
   %global cond;
   %let old=sales;
   %let cond=cases>0;
%mend conditn;
```

Here is the rest of the program:

```
%let new=inventry;

%macro name4;
   %let new=report;
   %let old=warehse;
   %conditn
   data &new;
      set &old;
      if &cond;
   run;
%mend name4;

%name4
```

Invoking NAME4 generates these statements:

```
data report;
   set sales;
   if cases>0;
run;
```

Suppose you want to put the SAS DATA step statements outside NAME4. In this case, all the macro variables must be global for the macro processor to resolve the references. You cannot add OLD to the %GLOBAL statement in CONDITN because the %LET statement in NAME4 has already created OLD as a local variable to NAME4 by the time CONDITN begins to execute. (You cannot use the %GLOBAL statement to make an existing local variable global.)

Thus, to make OLD global, use the %GLOBAL statement before the variable reference appears anywhere else, as shown here in the macro NAME5:

```
%let new=inventry;

%macro conditn;
   %global cond;
   %let old=sales;
   %let cond=cases>0;
%mend conditn;

   %macro name5;
      %global old;
      %let new=report;
      %let old=warehse;
```

```
      %conditn
   %mend name5;

%name5

data &new;
   set &old;
   if &cond;
run;
```

Now the %LET statement in NAME5 changes the value of the existing global variable OLD rather than creating OLD as a local variable. The SAS compiler sees the following statements:

```
data report;
   set sales;
   if cases>0;
run;
```

Creating Global Variables Based on the Value of Local Variables

To use a local variable such as a parameter outside a macro, use a %LET statement to assign the value to a global variable with a different name, as in this program:

```
%macro namels3(name,number);
   %local n;
   %global g_number;
   %let g_number=&number;
   %do n=1 %to &number;
      &name&n
   %end;
%mend namels3;
```

Now invoke the macro NAMELS3 in the following the program:

```
%let n=North State Industries;

proc print;
   var %namels3(dept,5);
   title "Quarterly Report for &n";
   footnote "Survey of &g_number Departments";
run;
```

The compiler sees the following statements:

```
proc print;
   var dept1 dept2 dept3 dept4 dept5;
   title "Quarterly Report for North State Industries";
   footnote "Survey of 5 Departments";
run;
```

Special Cases of Scope with the CALL SYMPUT Routine

Overview of CALL SYMPUT Routine

Most problems with CALL SYMPUT involve the lack of a precise step boundary between the CALL SYMPUT statement that creates the macro variable and the macro variable reference that uses that variable. (For more information, see "CALL SYMPUT Routine" on page 224.) However, a few special cases exist that involve the scope of a macro variable created by CALL SYMPUT. These cases are good examples of why you should always assign a scope to a variable before assigning a value rather than relying on SAS to do it for you.

Two rules control where CALL SYMPUT creates its variables:

1. CALL SYMPUT creates the macro variable in the current symbol table available while the DATA step is executing, provided that symbol table is not empty. If it is empty (contains no local macro variables), usually CALL SYMPUT creates the variable in the closest nonempty symbol table.
2. However, there are three cases where CALL SYMPUT creates the variable in the local symbol table, even if that symbol table is empty:
   - Beginning with SAS Version 8, if CALL SYMPUT is used after a PROC SQL, the variable will be created in a local symbol table.
   - If the macro variable SYSPBUFF is created at macro invocation time, the variable will be created in the local symbol table.
   - If the executing macro contains a computed %GOTO statement, the variable will be created in the local symbol table. A computed %GOTO statement is one that uses a label that contains an & or a % in it. That is, a computed %GOTO statement contains a macro variable reference or a macro call that produces a text expression. Here is an example of a computed %GOTO statement:

     ```
     %goto &home;
     ```

The symbol table that is currently available to a DATA step is the one that exists when SAS determines that the step is complete. (SAS considers a DATA step to be complete when it encounters a RUN statement, a semicolon after data lines, or the beginning of another step.)

In simplest terms, if an executing macro contains a computed %GOTO statement, or if the macro variable SYSPBUFF is created at macro invocation time, but the local symbol table is empty, CALL SYMPUT behaves as if the local symbol table was not empty, and creates a local macro variable.

You might find it helpful to use the %PUT statement with the _USER_ option to determine what symbol table the CALL SYMPUT routine has created the variable in.

Example Using CALL SYMPUT with Complete DATA Step and a Nonempty Local Symbol Table

Consider the following example, which contains a complete DATA step with a CALL SYMPUT statement inside a macro:

```
%macro env1(param1);
   data _null_;
      x = 'a token';
      call symput('myvar1',x);
   run;
%mend env1;

%env1(10)

data temp;
   y = "&myvar1";
run;
```

When you submit these statements, you receive an error message:

```
WARNING:  Apparent symbolic reference MYVAR1 not resolved.
```

This message appears because the DATA step is complete within the environment of ENV1 (that is, the RUN statement is within the macro) and because the local symbol table of ENV1 is not empty (it contains parameter PARAM1). Therefore, the CALL SYMPUT routine creates MYVAR1 as a local variable for ENV1, and the value is not available to the subsequent DATA step, which expects a global macro variable.

To see the scopes, add a %PUT statement with the _USER_ option to the macro, and a similar statement in open code. Now invoke the macro as before:

```
%macro env1(param1);
   data _null_;
      x = 'a token';
      call symput('myvar1',x);
   run;

   %put ** Inside the macro: **;
   %put _user_;
%mend env1;

%env1(10)

%put ** In open code: **;
%put _user_;

data temp;
   y = "&myvar1";   /* ERROR - MYVAR1 is not available in open code. */
run;
```

When the %PUT _USER_ statements execute, they write the following information to the SAS log:

```
** Inside the macro: **
ENV1    MYVAR1    a token
ENV1    PARAM1    10

** In open code: **
```

The MYVAR1 macro variable is created by CALL SYMPUT in the local ENV1 symbol table. The %PUT _USER_ statement in open code writes nothing to the SAS log, because no global macro variables are created.

The following figure shows all of the symbol tables in this example.

Figure 5.9 *The Symbol Tables with the CALL SYMPUT Routine Generating a Complete DATA Step*

Before ENV1 executes

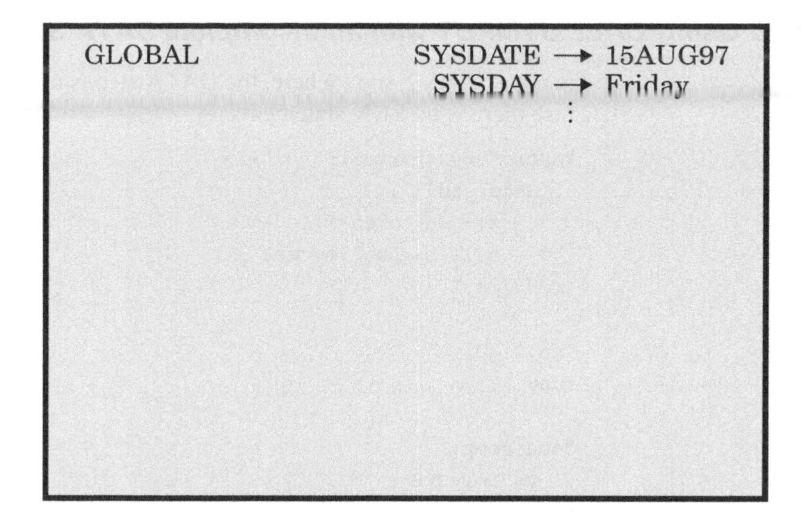

While ENV1 executes

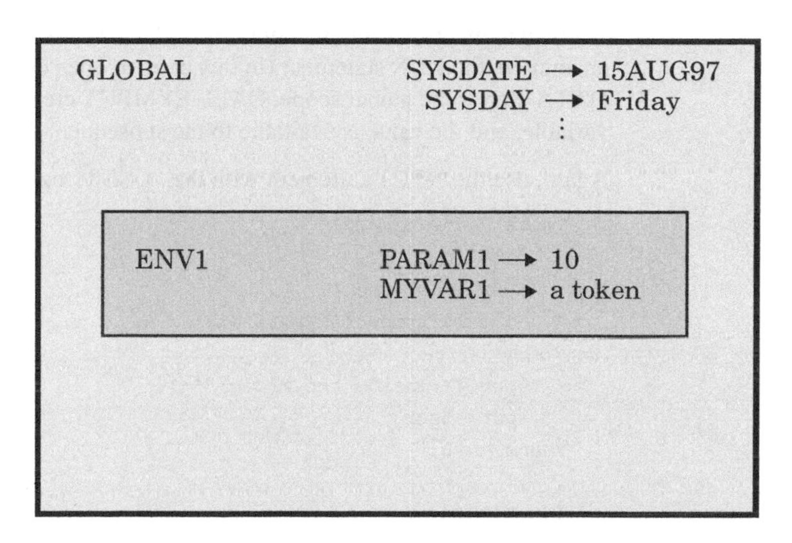

After ENV1 executes

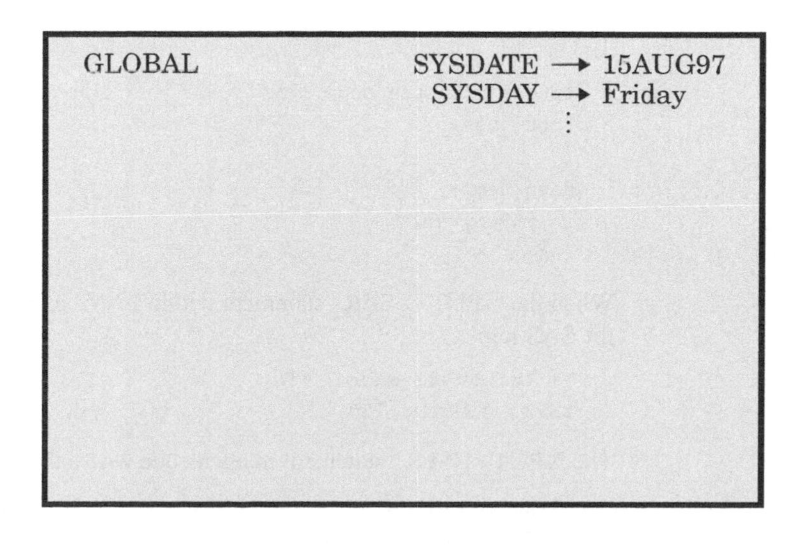

Example Using CALL SYMPUT with an Incomplete DATA Step

In the macro ENV2, shown here, the DATA step is not complete within the macro because there is no RUN statement:

```
%macro env2(param2);
   data _null_;
      x = 'a token';
      call symput('myvar2',x);
%mend env2;

%env2(20)
run;

data temp;
   y="&myvar2";
run;
```

These statements execute without errors. The DATA step is complete only when SAS encounters the RUN statement (in this case, in open code). Thus, the current scope of the DATA step is the global scope. CALL SYMPUT creates MYVAR2 as a global macro variable, and the value is available to the subsequent DATA step.

Again, use the %PUT statement with the _USER_ option to illustrate the scopes:

```
%macro env2(param2);
   data _null_;
      x = 'a token';
      call symput('myvar2',x);

   %put ** Inside the macro: **;
   %put _user_;
%mend env2;

%env2(20)

run;

%put ** In open code: **;
%put _user_;

data temp;
   y="&myvar2";
run;
```

When the %PUT _USER_ statement within ENV2 executes, it writes the following to the SAS log:

```
** Inside the macro: **
ENV2   PARAM2   20
```

The %PUT _USER_ statement in open code writes the following to the SAS log:

```
** In open code: **
GLOBAL   MYVAR2   a token
```

The following figure shows all the scopes in this example.

Figure 5.10 *The Symbol Tables with the CALL SYMPUT Routine Generating an Incomplete DATA Step*

Before ENV2 executes

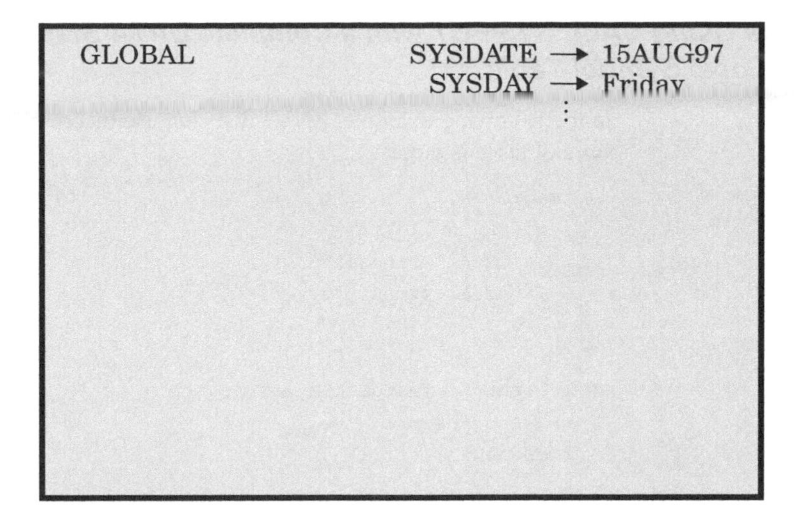

While ENV2 executes

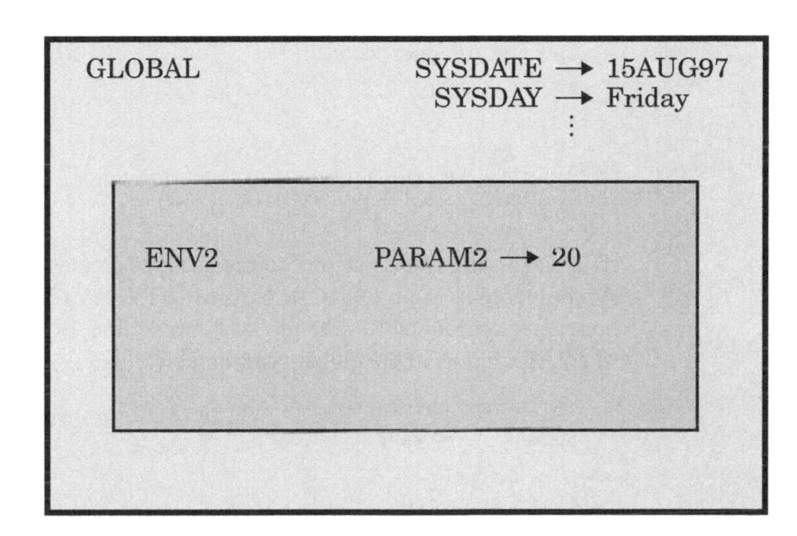

After ENV2 executes

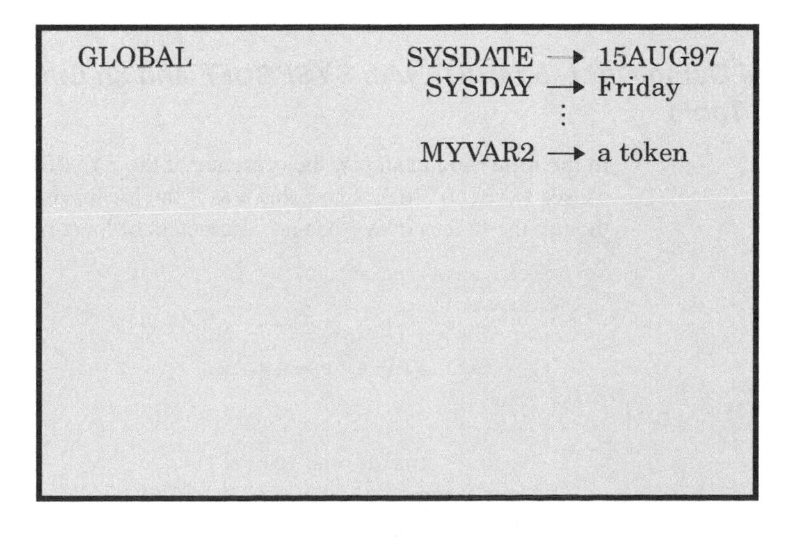

Example Using CALL SYMPUT with a Complete DATA Step and an Empty Local Symbol Table

In the following example, ENV3 does not use macro parameters. Therefore, its local symbol table is empty:

```
%macro env3;
   data _null_;
      x = 'a token';
      call symput('myvar3',x);
   run;

   %put ** Inside the macro: **;
   %put _user_;
%mend env3;

%env3

%put ** In open code: **;
%put _user_;

data temp;
   y="&myvar3";
run;
```

In this case, the DATA step is complete and executes within the macro, but the local symbol table is empty. So, CALL SYMPUT creates MYVAR3 in the closest available nonempty symbol table—the global symbol table. Both %PUT statements show that MYVAR3 exists in the global symbol table:

```
** Inside the macro: **
GLOBAL    MYVAR3   a token

** In open code: **
GLOBAL    MYVAR3   a token
```

Example Using CALL SYMPUT with SYSPBUFF and an Empty Local Symbol Table

In the following example, the presence of the SYSPBUFF automatic macro variable causes CALL SYMPUT to behave as if the local symbol table were not empty, even though the macro ENV4 has no parameters or local macro variables:

```
%macro env4 /parmbuff;
 data _null_;
      x = 'a token';
      call symput('myvar4',x);
   run;

   %put ** Inside the macro: **;
   %put _user_;
   %put &syspbuff;
%mend env4;

%env4
```

```
%put ** In open code: **;
%put _user_;
%put &syspbuff;

data temp;
   y="&myvar4";  /* ERROR - MYVAR4 is not available in open code */
run;
```

The presence of the /PARMBUFF specification causes the SYSPBUFF automatic macro variable to be created. So, when you call macro ENV4, CALL SYMPUT creates the macro variable MYVAR4 in the local symbol table (that is, in ENV4's), even though the macro ENV4 has no parameters and no local variables.

The results of the %PUT statements prove this—the score of MYVAR4 is listed as ENV4, and the reference to SYSPBUFF does not resolve in the open code %PUT statement because SYSPBUFF is local to ENV4:

```
** Inside the macro: **
b ENV4      MYVAR4    a token

** In open code: **
WARNING: Apparent symbolic reference SYSPBUFF not resolved.
```

For more information, see "SYSPBUFF Automatic Macro Variable" on page 210.

Chapter 6
Macro Expressions

Macro Expressions

There are three types of macro expressions: text, logical, and arithmetic. A *text expression* is any combination of text, macro variables, macro functions, or macro calls. Text expressions are resolved to generate text. Here are some examples of text expressions:

- &BEGIN
- %GETLINE
- &PREFIX.PART&SUFFIX
- %UPCASE(&ANSWER)

Logical expressions and *arithmetic expressions* are sequences of operators and operands forming sets of instructions that are evaluated to produce a result. An arithmetic expression contains an arithmetic operator. A logical expression contains a logical operator. The following table shows examples of simple arithmetic and logical expressions:

Table 6.1 *Arithmetic and Logical Expressions*

Arithmetic Expressions	Logical Expressions
1 + 2	&DAY = FRIDAY

Arithmetic Expressions	Logical Expressions
4 * 3	A < a
4 / 2	1 < &INDEX
00FFx - 003Ax	&START NE &END

Defining Arithmetic and Logical Expressions

Evaluating Arithmetic and Logical Expressions

You can use arithmetic and logical expressions in specific macro functions and statements. (See the following table.) The arithmetic and logical expressions in these functions and statements enable you to control the text generated by a macro when it is executed.

Table 6.2 *Macro Language Elements that Evaluate Arithmetic and Logical Expressions*

%DO*macro-variable=expression* %TO *expression*<%BY *expression*>;
%DO %UNTIL(*expression*);
%DO %WHILE(*expression*);
%EVAL (*expression*);
%IF *expression* %THEN *statement*;
%QSCAN(*argument,expression*<,*delimiters*>)
%QSUBSTR(*argument,expression*<,expression>)
%SCAN(*argument,expression*,<delimiters>)
%SUBSTR(*argument,expression*<,expression>)
%SYSEVALF(*expression,conversion-type*)

You can use text expressions to generate partial or complete arithmetic or logical expressions. The macro processor resolves text expressions before it evaluates the arithmetic or logical expressions. For example, when you submit the following statements, the macro processor resolves the macro variables &A, &B, and &OPERATOR in the %EVAL function, before it evaluates the expression 2 + 5:

```
%let A=2;
%let B=5;
%let operator=+;
```

```
%put The result of &A &operator &B is %eval(&A &operator
&B).;
```

When you submit these statements, the %PUT statement writes the following to the log:

```
The result of 2 + 5 is 7.
```

Operands and Operators

Operands in arithmetic or logical expressions are always text. However, an operand that represents a number can be temporarily converted to a numeric value when an expression is evaluated. By default, the macro processor uses integer arithmetic, and only integers and hexadecimal values that represent integers can be converted to a numeric value. Operands that contain a period character (for example 1.0) are not converted. The exception is the %SYSEVALF function. It interprets a period character in its argument as a decimal point and converts the operand to a floating-point value on your operating system.

Note: The values of numeric expressions are restricted to the range of –2**64 to 2**64–1.

Operators in macro expressions are a subset of the operators in the DATA step (Table 6.3 on page 73). However, in the macro language, there is no MAX or MIN operator, and it does not recognize ':', as does the DATA step. The order in which operations are performed when an expression is evaluated is the same in the macro language as in the DATA step. Operations within parentheses are performed first.

Note: Expressions in which comparison operators surround a macro expression, as in 10<&X<20, might be the equivalent of a DATA step compound expression (depending on the expression resolution). To be safe, specify the connecting operator, as in the expression 10<&X AND &X<20.

Table 6.3 *Macro Language Operators*

Operator	Mnemonic	Precedence	Definition	Example
**		1	exponentiation	2**4
+		2	positive prefix	+(A+B)
-		2	negative prefix	-(A+B)
¬^~	NOT	3	logical not*	NOT A
*		4	multiplication	A*B
/		4	division	A/B
+		5	addition	A+B
-		5	subtraction	A-B
<	LT	6	less than	A<B

Operator	Mnemonic	Precedence	Definition	Example
<=	LE	6	less than or equal	A<=B
=	EQ	6	equal	A=B
#	IN	6	equal to one of a list**	A#B C D E
¬= ^= ~=	NE	6	not equal*	A NE B
>	GT	6	greater than	A>B
>=	GE	6	greater than or equal	A>=B
&	AND	7	logical and	A=B & C=D
|	OR	8	logical or	A=B | C=D

*The symbol to use depends on your keyboard.

** The default delimiter for list elements is a blank. See "MINDELIMITER= System Option" on page 343 for more information.

** Before using the IN (#) operator, see "MINOPERATOR System Option" on page 345.

** When you use the IN operator, both operands must contain a value. If the operand contains a null value, an error is generated.

CAUTION:
Integer expressions that contain exponential, multiplication, or division operators and that use or compute values outside the range -9,007,199,254,740,992 to 9,007,199,254,740,992 might get inexact results.

How the Macro Processor Evaluates Arithmetic Expressions

Evaluating Numeric Operands

The macro facility is a string handling facility. However, in specific situations, the macro processor can evaluate operands that represent numbers as numeric values. When the macro processor evaluates an expression that contains an arithmetic operator and operands that represent numbers, it temporarily converts the operands to numeric values and performs the integer arithmetic operation. The result of the evaluation is text.

By default, arithmetic evaluation in most macro statements and functions is performed with integer arithmetic. The exception is the %SYSEVALF function. See "Evaluating Floating-Point Operands" on page 75 for more information. The following macro statements illustrate integer arithmetic evaluation:

```
%let a=%eval(1+2);
%let b=%eval(10*3);
```

```
%let c=%eval(4/2);
%let i=%eval(5/3);
%put The value of a is &a;
%put The value of b is &b;
%put The value of c is &c;
%put The value of I is &i;
```

When you submit these statements, the following messages appear in the log:

```
The value of a is 3
The value of b is 30
The value of c is 2
The value of I is 1
```

Notice the result of the last statement. If you perform division on integers that would ordinarily result in a fraction, integer arithmetic discards the fractional part.

When the macro processor evaluates an integer arithmetic expression that contains a character operand, it generates an error. Only operands that contain characters that represent integers or hexadecimal values are converted to numeric values. The following statement shows an incorrect usage:

```
%let d=%eval(10.0+20.0);    /*INCORRECT*/
```

Because the %EVAL function supports only integer arithmetic, the macro processor does not convert a value containing a period character to a number, and the operands are evaluated as character operands. This statement produces the following error message:

```
ERROR: A character operand was found in the %EVAL function or %IF
condition where a numeric operand is required. The condition was:
10.0+20.0
```

Evaluating Floating-Point Operands

The %SYSEVALF function evaluates arithmetic expressions with operands that represent floating-point values. For example, the following expressions in the %SYSEVALF function are evaluated using floating-point arithmetic:

```
%let a=%sysevalf(10.0*3.0);
%let b=%sysevalf(10.5+20.8);
%let c=%sysevalf(5/3);
%put 10.0*3.0 = &a;
%put 10.5+20.8 = &b;
%put 5/3 = &c;
```

The %PUT statements display the following messages in the log:

```
10.0*3.0 = 30
10.5+20.8 = 31.3
5/3 = 1.6666666667
```

When the %SYSEVALF function evaluates arithmetic expressions, it temporarily converts the operands that represent numbers to floating-point values. The result of the evaluation can represent a floating-point value, but as in integer arithmetic expressions, the result is always text.

The %SYSEVALF function provides conversion type specifications: BOOLEAN, INTEGER, CEIL, and FLOOR. For example, the following %PUT statements return 1, 2, 3, and 2 respectively:

```
%let a=2.5;
%put %sysevalf(&a,boolean);
```

```
%put %sysevalf(&a,integer);
%put %sysevalf(&a,ceil);
%put %sysevalf(&a,floor);
```

These conversion types tailor the value returned by %SYSEVALF so that it can be used in other macro expressions that require integer or Boolean values.

CAUTION:

Specify a conversion type for the %SYSEVALF function. If you use the %SYSEVALF function in macro expressions or assign its results to macro variables that are used in other macro expressions, then errors or unexpected results might occur if the %SYSEVALF function returns missing or floating-point values. To prevent errors, specify a conversion type that returns a value compatible with other macro expressions. See "%SYSEVALF Function" on page 263 for more information about using conversion types.

How the Macro Processor Evaluates Logical Expressions

Comparing Numeric Operands in Logical Expressions

A logical, or Boolean, expression returns a value that is evaluated as true or false. In the macro language, any numeric value other than 0 is true and a value of 0 is false.

When the macro processor evaluates logical expressions that contain operands that represent numbers, it converts the characters temporarily to numeric values. To illustrate how the macro processor evaluates logical expressions with numeric operands, consider the following macro definition:

```
%macro compnum(first,second);
   %if &first>&second %then %put &first is greater than &second;
   %else %if &first=&second %then %put &first equals &second;
   %else %put &first is less than &second;
%mend compnum;
```

Invoke the COMPNUM macro with these values:

```
%compnum(1,2)
%compnum(-1,0)
```

The following results are displayed in the log:

```
1 is less than 2
-1 is less than 0
```

The results show that the operands in the logical expressions were evaluated as numeric values.

Comparing Floating-Point or Missing Values

You must use the %SYSEVALF function to evaluate logical expressions containing floating-point or missing values. To illustrate comparisons with floating-point and missing values, consider the following macro that compares parameters passed to it with the %SYSEVALF function and places the result in the log:

```
%macro compflt(first,second);
   %if %sysevalf(&first>&second) %then %put &first is greater than
&second;
   %else %if  %sysevalf(&first=&second) %then %put &first equals
&second;
   %else %put &first is less than &second;
%mend compflt;
```

Invoke the COMPFLT macro with these values:

```
%compflt (1.2,.9)
%compflt (-.1,.)
%compflt (0,.)
```

The following values are written in the log:

```
1.2 is greater than .9
-.1 is greater than .
0 is greater than .
```

The results show that the %SYSEVALF function evaluated the floating-point and missing values.

Comparing Character Operands in Logical Expressions

To illustrate how the macro processor evaluates logical expressions, consider the COMPCHAR macro. Invoking the COMPCHAR macro compares the values passed as parameters and places the result in the log.

```
%macro compchar(first,second);
   %if &first>&second %then %put &first comes after &second;
   %else %put &first comes before &second;
%mend compchar;
```

Invoke the macro COMPCHAR with these values:

```
%compchar(a,b)
%compchar(.,1)
%compchar(Z,E)
```

The following results are printed in the log:

```
a comes before b
. comes before 1
Z comes after E
```

When the macro processor evaluates expressions with character operands, it uses the sort sequence of the host operating system for the comparison. The comparisons in these examples work with both EBCDIC and ASCII sort sequences.

A special case of a character operand is an operand that looks numeric but contains a period character. If you use an operand with a period character in an expression, both operands are compared as character values. This can lead to unexpected results. So that you can understand and better anticipate results, look at the following examples.

Invoke the COMPNUM macro with these values:

```
%compnum(10,2.0)
```

The following values are written to the log:

```
10 is less than 2.0
```

Because the %IF-THEN statement in the COMPNUM macro uses integer evaluation, it does not convert the operands with decimal points to numeric values. The operands are compared as character strings using the host sort sequence, which is the comparison of characters with smallest-to-largest values. For example, lowercase letters might have smaller values than uppercase, and uppercase letters might have smaller values than digits.

CAUTION:

The host sort sequence determines comparison results. If you use a macro definition on more than one operating system, comparison results might differ because the sort sequence of one host operating system might differ from the other system. See the Chapter 48, "SORT Procedure" in *Base SAS Procedures Guide* for more information about host sort sequences.

Chapter 7

Macro Quoting

Macro Quoting

Masking Special Characters and Mnemonics

The macro language is a character-based language. Even variables that appear to be numeric are generally treated as character variables (except during expression evaluation). Therefore, the macro processor enables you to generate all sorts of special characters as text. But because the macro language includes some of the same special characters, an ambiguity often arises. The macro processor must know whether to interpret a particular special character (for example, a semicolon or % sign) or a mnemonic (for example, GE or AND) as text or as a symbol in the macro language. Macro quoting functions resolve these ambiguities by masking the significance of special characters so that the macro processor does not misinterpret them.

The following special characters and mnemonics might require masking when they appear in text strings:

Table 7.1 *Special Characters and Mnemonics*

<table>
<tr><td>blank</td><td>)</td><td>=</td><td>LT</td></tr>
<tr><td>;</td><td>(</td><td>|</td><td>GE</td></tr>
<tr><td>¬</td><td>+</td><td>AND</td><td>GT</td></tr>
<tr><td>^</td><td>—</td><td>OR</td><td>IN</td></tr>
<tr><td>~</td><td>*</td><td>NOT</td><td>%</td></tr>
<tr><td>, (comma)</td><td>/</td><td>EQ</td><td>&</td></tr>
<tr><td>'</td><td><</td><td>NE</td><td>#</td></tr>
<tr><td>“</td><td>></td><td>LE</td><td></td></tr>
</table>

Understanding Why Macro Quoting Is Necessary

Macro quoting functions tell the macro processor to interpret special characters and mnemonics as text rather than as part of the macro language. If you did not use a macro quoting function to mask the special characters, the macro processor or the rest of SAS might give the character a meaning that you did not intend. Here are some examples of the types of ambiguities that can arise when text strings contain special characters and mnemonics:

- Is `%sign` a call to the macro SIGN or a phrase “percent sign”?
- Is OR the mnemonic Boolean operator or the abbreviation for Oregon?
- Is the quotation mark in O'Malley an unbalanced single quotation mark or just part of the name?

- Is Boys&Girls a reference to the macro variable &GIRLS or a group of children?
- Is GE the mnemonic for "greater than or equal" or is it short for General Electric?
- Which statement does a semicolon end?
- Does a comma separate parameters, or is it part of the value of one of the parameters?

Macro quoting functions enable you to clearly indicate to the macro processor how it is to interpret special characters and mnemonics.

Here is an example, using the simplest macro quoting function, %STR. Suppose you want to assign a PROC PRINT statement and a RUN statement to the macro variable PRINT. Here is the erroneous statement:

```
%let print=proc print; run;;  /* undesirable results */
```

This code is ambiguous. Are the semicolons that follow PRINT and RUN part of the value of the macro variable PRINT, or does one of them end the %LET statement? If you do not tell the macro processor what to do, it interprets the semicolon after PRINT as the end of the %LET statement. So the value of the PRINT macro variable would be the following:

```
proc print
```

The rest of the characters (RUN;;) would be simply the next part of the program.

To avoid the ambiguity and correctly assign the value of PRINT, you must mask the semicolons with the macro quoting function %STR, as follows:

```
%let print=%str(proc print; run;);
```

Overview of Macro Quoting Functions

The following macro quoting functions are most commonly used:

- %STR and %NRSTR
- %BQUOTE and %NRBQUOTE
- %SUPERQ

For the paired macro quoting functions, the function beginning with NR affects the same category of special characters that are masked by the plain macro quoting function as well as ampersands and percent signs. In effect, the NR functions prevent macro and macro variable resolution. To help you remember which does which, try associating the NR in the macro quoting function names with the words "not resolved" — that is, macros and macro variables are not resolved when you use these functions.

The macro quoting functions with B in their names are useful for macro quoting unmatched quotation marks and parentheses. To help you remember the B, try associating B with "by itself".

The %SUPERQ macro quoting function is unlike the other macro quoting functions in that it does not have a mate and works differently. See "%SUPERQ Function" on page 259 for more information.

The macro quoting functions can also be divided into two types, depending on when they take effect:

compilation functions
: cause the macro processor to interpret special characters as text in a macro program statement in open code or while compiling (constructing) a macro. The %STR and

%NRSTR functions are compilation functions. For more information, see "%STR and %NRSTR Functions" on page 254.

execution functions
: cause the macro processor to treat special characters that result from resolving a macro expression as text (such as a macro variable reference, a macro invocation, or the argument of an %EVAL function). They are called execution functions because resolution occurs during macro execution or during execution of a macro program statement in open code. The macro processor resolves the expression as far as possible, issues any warning messages for macro variable references or macro invocations that it cannot resolve, and quotes the result. The %BQUOTE and %NRBQUOTE functions are execution functions. For more information, see "%BQUOTE and %NRBQUOTE Functions" on page 242.

The %SUPERQ function takes as its argument a macro variable name (or a macro expression that yields a macro variable name). The argument must not be a reference to the macro variable whose value you are masking. That is, do not include the & before the name.

Note: Two other execution macro quoting functions exist: %QUOTE and %NRQUOTE. They are useful for unique macro quoting needs and for compatibility with older macro applications. For more information, see "%QUOTE and %NRQUOTE Functions" on page 248.

Passing Parameters That Contain Special Characters and Mnemonics

Using an execution macro quoting function in the macro definition is the simplest and best way to have the macro processor accept resolved values that might contain special characters. However, if you discover that you need to pass parameter values such as `or` when a macro has not been defined with an execution macro quoting function, you can do so by masking the value in the macro invocation. The logic of the process is as follows:

1. When you mask a special character with a macro quoting function, it remains masked as long as it is within the macro facility (unless you use the "%UNQUOTE Function" on page 274).
2. The macro processor constructs the complete macro invocation before beginning to execute the macro.
3. Therefore, you can mask the value in the invocation with the %STR function. Although the masking is not needed when the macro processor is constructing the invocation, the value is already masked by a macro quoting function when macro execution begins and therefore does not cause problems during macro execution.

For example, suppose a macro named ORDERX does not use the %BQUOTE function. You can pass the value `or` to the ORDERX macro with the following invocation:

```
%orderx(%str(or))
```

However, placing the macro quoting function in the macro definition makes the macro much easier for you to invoke.

Deciding When to Use a Macro Quoting Function and Which Function to Use

Use a macro quoting function any time you want to assign to a macro variable a special character that could be interpreted as part of the macro language. The following table describes the special characters to mask when used as part of a text string and which macro quoting functions are useful in each situation.

Table 7.2 *Special Characters and Macro Quoting Guidelines*

<table>
<tr><th>Special Character</th><th>Must Be Masked</th><th>Quoted by All Macro Quoting Functions?</th><th>Remarks</th></tr>
<tr><td>+-*/<>=^|¬ ~ #
LE LT EQ NE
GE GT AND OR
NOT IN</td><td>To prevent it from being treated as an operator in the argument of an %EVAL function</td><td>Yes</td><td>AND, OR, IN, and NOT need to be masked because they are interpreted as mnemonic operators by an %EVAL and by %SYSEVALF.</td></tr>
<tr><td>blank</td><td>To maintain, rather than ignore, a leading, trailing, or isolated blank</td><td>Yes</td><td></td></tr>
<tr><td>;</td><td>To prevent a macro program statement from ending prematurely</td><td>Yes</td><td></td></tr>
<tr><td>, (comma)</td><td>To prevent it from indicating a new function argument, parameter, or parameter value</td><td>Yes</td><td></td></tr>
</table>

Special Character	Must Be Masked	Quoted by All Macro Quoting Functions?	Remarks
' " ( )	If it might be unmatched	No	Arguments that might contain quotation marks and parentheses should be masked with a macro quoting function so that the macro facility interprets the single and double quotation marks and parentheses as text rather than macro language symbols or possibly unmatched quotation marks or parentheses for the SAS language. With %STR, %NRSTR, %QUOTE, and %NRQUOTE, unmatched quotation marks and parentheses must be marked with a % sign. You do not have to mark unmatched symbols in the arguments of %BQUOTE, %NRBQUOTE, and %SUPERQ.
%*name* &*name*	(Depends on what the expression might resolve to)	No	%NRSTR, %NRBQUOTE, and %NRQUOTE mask these patterns. To use %SUPERQ with a macro variable, omit the ampersand from *name*.

The macro facility allows you as much flexibility as possible in designing your macros. You need to mask a special character with a macro quoting function only when the macro processor would otherwise interpret the special character as part of the macro language rather than as text. For example, in this statement you must use a macro quoting function to mask the first two semicolons to make them part of the text:

```
%let p=%str(proc print; run;);
```

However, in the macro PR, shown here, you do not need to use a macro quoting function to mask the semicolons after PRINT and RUN:

```
%macro pr(start);
   %if &start=yes %then
      %do;
```

```
            %put proc print requested;
            proc print;
            run;
         %end;
   %mend pm;
```

Because the macro processor does not expect a semicolon within the %DO group, the semicolons after PRINT and RUN are not ambiguous, and they are interpreted as text.

Although it is not possible to give a series of rules that cover every situation, the following sections describe how to use each macro quoting function. Table 7.6 on page 94 provides a summary of the various characters that might need masking and of which macro quoting function is useful in each situation.

Note: You can also perform the inverse of a macro quoting function — that is, remove the tokenization provided by macro quoting functions. For an example of when the %UNQUOTE function is useful, see "Unquoting Text " on page 95.

%STR and %NRSTR Functions

Using %STR and %NRSTR Functions

If a special character or mnemonic affects the way the macro processor constructs macro program statements, you must mask the item during macro compilation (or during the compilation of a macro program statement in open code) by using either the %STR or %NRSTR macro quoting functions.

These macro quoting functions mask the following special characters and mnemonics:

Table 7.3 *Special Characters Masked by the %STR and %NRSTR Functions*

blank	)	=	NE	
;	(	|	LE	
¬	+	#	LT	
^	—	AND	GE	
~	*	OR	GT	
, (comma)	/	NOT		
'	<	IN		
"	>	EQ		

In addition to these special characters and mnemonics, %NRSTR masks & and %.

Note: If an unmatched single or double quotation mark or an open or close parenthesis is used with %STR or %NRSTR, these characters must be preceded by a percent sign (%).

When you use %STR or %NRSTR, the macro processor does not receive these functions and their arguments when it executes a macro. It receives only the results of these functions because these functions work when a macro compiles. By the time the macro executes, the string is already masked by a macro quoting function. Therefore, %STR and %NRSTR are useful for masking strings that are constants, such as sections of SAS code. In particular, %NRSTR is a good choice for masking strings that contain % and & signs. However, these functions are not so useful for masking strings that contain references to macro variables because it is possible that the macro variable could resolve to a value not quotable by %STR or %NRSTR. For example, the string could contain an unmarked, unmatched open parenthesis.

Using Unmatched Quotation Marks and Parentheses with %STR and %NRSTR

If the argument to %STR or %NRSTR contains an unmatched single or double quotation mark or an unmatched open or close parenthesis, precede each of these characters with a % sign. The following table shows some examples of this technique.

Table 7.4 *Examples of Marking Unmatched Quotation Marks and Parentheses with %STR and %NRSTR*

Notation	Description	Example	Quoted Value Stored
%'	unmatched single quotation mark	`%let myvar= %str(a%');`	`a'`
%"	unmatched double quotation mark	`%let myvar= %str(title %"first);`	`title "first`
%(	unmatched open parenthesis	`%let myvar= %str (log %(12);`	`log(12`
%)	unmatched close parenthesis	`%let myvar= %str (345%));`	`345)`

Using % Signs with %STR

In general, if you want to mask a % sign with a macro quoting function at compilation, use %NRSTR. There is one case where you can use %STR to mask a % sign: when the % sign does not have any text following it that could be construed by the macro processor as a macro name. The % sign must be marked by another % sign. Here are some examples.

Table 7.5 Examples of Masking % Signs with %STR

Notation	Description	Example	Quoted Value Stored
'%'	% sign before a matched single quotation mark	`%let myvar= %str('%');`	'%'
%%%'	% sign before an unmatched single quotation mark	`%let myvar= %str(%%%');`	`%'`
""%%	% sign after a matched double quotation mark	`%let myvar= %str(""%%);`	`""%`
%%%%	two % signs in a row	`%let myvar= %str(%%%%);`	`%%`

Examples Using %STR

The %STR function in the following %LET statement prevents the semicolon after PROC PRINT from being interpreted as the ending semicolon for the %LET statement:

```
%let printit=%str(proc print; run;);
```

As a more complex example, the macro KEEPIT1 shows how the %STR function works in a macro definition:

```
%macro keepit1(size);
   %if &size=big %then %put %str(keep city _numeric_;);
   %else %put %str(keep city;);
%mend keepit1;
```

Call the macro as follows:

```
%keepit1(big)
```

This code produces the following statement:

```
keep city _numeric_;
```

When you use the %STR function in the %IF-%THEN statement, the macro processor interprets the first semicolon after the word %THEN as text. The second semicolon ends the %THEN statement, and the %ELSE statement immediately follows the %THEN statement. Thus, the macro processor compiles the statements as you intended. However, if you omit the %STR function, the macro processor interprets the first semicolon after the word %THEN as the end of the %THEN clause. The next semicolon as constant text. Because only a %THEN clause can precede a %ELSE clause, the semicolon as constant text causes the macro processor to issue an error message and not compile the macro.

In the %ELSE statement, the %STR function causes the macro processor to treat the first semicolon in the statement as text and the second one as the end of the %ELSE clause. Therefore, the semicolon that ends the KEEP statement is part of the conditional execution. If you omit the %STR function, the first semicolon ends the %ELSE clause and the second semicolon is outside the conditional execution. It is generated as text each time the macro executes. (In this example, the placement of the semicolon does not affect the SAS code.) Again, using %STR causes the macro KEEPIT1 to compile as you intended.

Here is an example that uses %STR to mask a string that contains an unmatched single quotation mark. Note the use of the % sign before the quotation mark:

```
%let innocent=%str(I didn%'t do it!);
```

Examples Using %NRSTR

Suppose you want the name (not the value) of a macro variable to be printed by the %PUT statement. To do so, you must use the %NRSTR function to mask the & and prevent the resolution of the macro variable, as in the following example:

```
%macro example;
   %local myvar;
   %let myvar=abc;
   %put %nrstr(The string &myvar appears in log output,);
   %put instead of the variable value.;
%mend example;

%example
```

This code writes the following text to the SAS log:

```
The string &myvar appears in log output,
instead of the variable value.
```

If you did not use the %NRSTR function or if you used %STR, the following undesired output would appear in the SAS log:

```
The string abc appears in log output,
instead of the variable value.
```

The %NRSTR function prevents the & from triggering macro variable resolution.

The %NRSTR function is also useful when the macro definition contains patterns that the macro processor would ordinarily recognize as macro variable references, as in the following program:

```
%macro credits(d=%nrstr(Mary&Stacy&Joan Ltd.));
   footnote "Designed by &d";
%mend credits;
```

Using %NRSTR causes the macro processor to treat &STACY and &JOAN simply as part of the text in the value of D; the macro processor does not issue warning messages for unresolvable macro variable references. Suppose you invoke the macro CREDITS with the default value of D, as follows:

```
%credits()
```

Submitting this program generates the following FOOTNOTE statement:

```
footnote "Designed by Mary&Stacy&Joan Ltd.";
```

If you omit the %NRSTR function, the macro processor attempts to resolve the references &STACY and &JOAN as part of the resolution of &D in the FOOTNOTE statement. The macro processor issues these warning messages (assuming the SERROR system option, described in , on page 329 is active) because no such macro variables exist:

```
WARNING: Apparent symbolic reference STACY not resolved.
WARNING: Apparent symbolic reference JOAN not resolved.
```

Here is a final example of using %NRSTR. Suppose you wanted to have a text string include the name of a macro function: **`This is the result of %NRSTR`**. Here is the program:

```
%put This is the result of %nrstr(%nrstr);
```

You must use %NRSTR to mask the % sign at compilation, so the macro processor does not try to invoke %NRSTR a second time. If you did not use %NRSTR to mask the string **`%nrstr`**, the macro processor would complain about a missing open parenthesis for the function.

%BQUOTE and %NRBQUOTE Functions

Using %BQUOTE and %NRBQUOTE Functions

%BQUOTE and %NRBQUOTE mask values during execution of a macro or a macro language statement in open code. These functions instruct the macro processor to resolve a macro expression as far as possible and mask the result, issuing any warning messages for macro variable references or macro invocations that it cannot resolve. These functions mask all the characters that %STR and %NRSTR mask with the addition of unmarked percent signs; unmatched, unmarked single and double quotation marks; and unmatched, unmarked opening and closing parentheses. That means that you do not have to precede an unmatched quotation mark with a % sign, as you must when using %STR and %NRSTR.

The %BQUOTE function treats all parentheses and quotation marks produced by resolving macro variable references or macro calls as special characters to be masked at execution time. (It does not mask parentheses or quotation marks that are in the argument at compile time.) Therefore, it does not matter whether quotation marks and parentheses in the resolved value are matched; each one is masked individually.

The %NRBQUOTE function is useful when you want a value to be resolved when first encountered, if possible, but you do not want any ampersands or percent signs in the result to be interpreted as operators by an %EVAL function.

If the argument of the %NRBQUOTE function contains an unresolvable macro variable reference or macro invocation, the macro processor issues a warning message before it masks the ampersand or percent sign (assuming the SERROR or MERROR system option, described in "System Options for Macros" on page 329 is in effect). To suppress the message for unresolved macro variables, use the %SUPERQ function (discussed later in this section) instead.

Because the %BQUOTE and %NRBQUOTE functions operate during execution and are more flexible than %STR and %NRSTR, %BQUOTE and %NRBQOUTE are good choices for masking strings that contain macro variable references.

Examples Using %BQUOTE

In the following statement, the %IF-%THEN statement uses %BQUOTE to prevent an error if the macro variable STATE resolves to OR (for Oregon), which the macro processor would interpret as the logical operator OR otherwise:

```
%if %bquote(&state)=%str(OR) %then %put Oregon Dept. of
Revenue;
```

Note: This example works if you use %STR, but it is not robust or good programming practice. Because you cannot guarantee what &STATE is going to resolve to, you need to use %BQUOTE to mask the resolution of the macro variable at execution time, not the name of the variable itself at compile time.

In the following example, a DATA step creates a character value containing a single quotation mark and assigns that value to a macro variable. The macro READIT then uses the %BQUOTE function to enable a %IF condition to accept the unmatched single quotation mark:

```
data test;
   store="Susan's Office Supplies";
   call symput('s',store);
run;

%macro readit;
   %if %bquote(&s) ne %then %put *** valid ***;
   %else %put *** null value ***;
%mend readit;

%readit
```

When you assign the value **`Susan's Office Supplies`**to STORE in the DATA step, enclosing the character string in double quotation marks enables you to use an unmatched single quotation mark in the string. SAS stores the value of STORE:

```
Susan's Office Supplies
```

The CALL SYMPUT routine assigns that value (containing an unmatched single quotation mark) as the value of the macro variable S. If you do not use the %BQUOTE function when you reference S in the macro READIT, the macro processor issues an error message for an invalid operand in the %IF condition.

When you submit the code, the following is written to the SAS log:

```
*** valid ***
```

Referring to Already Quoted Variables

Items that have been masked by a macro quoting function, such as the value of WHOSE in the following program, remain masked as long as the item is being used by the macro processor. When you use the value of WHOSE later in a macro program statement, you do not need to mask the reference again.

```
/* Use %STR to mask the constant, and use a  % sign to mark */
/* the unmatched single quotation mark. */
%let whose=%str(John%'s);

/* You don't need to mask the macro reference, because it was */
/* masked in the %LET statement, and remains masked. */
%put *** This coat is &whose ***;
```

Here is the output from the %PUT statement that is written to the SAS log:

```
*** This coat is John's ***
```

Deciding How Much Text to Mask with a Macro Quoting Function

In each of the following statements, the macro processor treats the masked semicolons as text:

```
%let p=%str(proc print; run;);
%let p=proc %str(print;) %str(run;);
%let p=proc print%str(;) run%str(;);
```

The value of P is the same in each case:

```
proc print; run;
```

The results of the three %LET statements are the same because when you mask text with a macro quoting function, the macro processor quotes only the items that the function recognizes. Other text enclosed in the function remains unchanged. Therefore, the third %LET statement is the minimalist approach to macro quoting. However, masking large blocks of text with a macro quoting function is harmless and actually results in code that is much easier to read (such as the first %LET statement).

%SUPERQ Function

Using %SUPERQ

The %SUPERQ function locates the macro variable named in its argument and quotes the value of that macro variable without permitting any resolution to occur. It masks all items that might require macro quoting at macro execution. Because %SUPERQ does not attempt any resolution of its argument, the macro processor does not issue any warning messages that a macro variable reference or a macro invocation has not been resolved. Therefore, even when the %NRBQUOTE function enables the program to work correctly, you can use the %SUPERQ function to eliminate unwanted warning messages from the SAS log. %SUPERQ takes as its argument either a macro variable name without an ampersand or a text expression that yields a macro variable name.

%SUPERQ retrieves the value of a macro variable from the macro symbol table and quotes it immediately, preventing the macro processor from making any attempt to resolve anything that might occur in the resolved value. For example, if the macro variable CORPNAME resolves to **`Smith&Jones`**, using %SUPERQ prevents the macro processor from attempting to further resolve **`&Jones`**. This %LET statement successfully assigns the value **`Smith&Jones`** to TESTVAR:

```
%let testvar=%superq(corpname);
```

Examples Using %SUPERQ

This example shows how the %SUPERQ function affects two macro invocations, one for a macro that has been defined and one for an undefined macro:

```
%window ask
 #5 @5 'Enter two values:'
```

```
 #5 @24 val 15 attr=underline;

%macro a;
   %put *** This is a. ***;
%mend a;

%macro test;
   %display ask;
   %put *** %superq(val) ***;     /* Note absence of ampersand */
%mend test;
```

Suppose you invoke the macro TEST and respond to the prompt as shown:

```
%test
Enter the following:  %a %x
```

The %PUT statement simply writes the following line:

```
*** %a %x ***
```

It does not invoke the macro A, and it does not issue a warning message that %X was not resolved. The following two examples compare the %SUPERQ function with other macro quoting functions.

Using the %SUPERQ Function to Prevent Warning Messages

The sections about the %NRBQUOTE function show that it causes the macro processor to attempt to resolve the patterns *&name* and *%name* the first time it encounters them during macro execution. If the macro processor cannot resolve them, it quotes the ampersand or percent sign so that later uses of the value do not cause the macro processor to recognize them. However, if the MERROR or SERROR option is in effect, the macro processor issues a warning message that the reference or invocation was not resolved.

The macro FIRMS3, shown here, shows how the %SUPERQ function can prevent unwanted warning messages:

```
%window ask
    #5 @5 'Enter the name of the company:'
    #5 @37 val 50 attr=underline;

%macro firms3;
    %global code;
    %display ask;
        %let name=%superq(val);
           %if &name ne %then %let code=valid;
           %else %let code=invalid;
           %put *** &name is &code ***;
%mend firms3;

%firms3
```

Suppose you invoke the macro FIRMS3 twice and respond with the following companies:

```
A&A Autos
Santos&D'Amato
```

After the macro executes, the following is written to the SAS log:

```
*** A&A Autos is valid ***
*** Santos&D'Amato is valid ***
```

Using the %SUPERQ Function to Enter Macro Keywords

Suppose you create an online training system in which users can enter problems and questions that another macro prints for you later. The user's response to a %WINDOW statement is assigned to a local macro variable and then to a global macro variable. Because the user is asking questions about macros, he or she might enter all sorts of macro variable references and macro calls as examples of problems, as well as unmatched, unmarked quotation marks and parentheses. If you mask the response with %BQUOTE, you have used a few %PUT statements to warn the user about responses that cause problems. If you the %SUPERQ function, you need fewer instructions. The macros ASK1 and ASK2 show how the macro code becomes simpler as you change macro quoting functions.

The macro ASK1, below, shows how the macro looks when you use the %BQUOTE function:

```
%window ask
    #5 @5 'Describe the problem.'
    #6 @5 'Do not use macro language keywords, macro calls,'
    #7 @5 'or macro variable references.'
    #9 @5 'Enter /// when you are finished.'
   #11 @5 val 100 attr=underline;

%macro ask1;
    %global myprob;
   %local temp;

   %do %until(%bquote(&val) eq %str(///));
       %display ask;
       %let temp=&temp %bquote(&val);
   %end;
    %let myprob=&temp
%mend ask1;
```

The macro ASK1 does not include a warning about unmatched quotation marks and parentheses. You can invoke the macro ASK1 and enter a problem:

```
%ask1

Try entering:
Why did my macro not run when I called it? (It had three
parameters, but I wasn't using any of them.)
It ran after I submitted the next statement.
///
```

Notice that both the first and second lines of the response contain an unmatched, unmarked quotation mark and parenthesis. %BQUOTE can handle these characters during execution.

The macro ASK2, shown here, modifies the macro ASK1 by using the %SUPERQ function. Now the %WINDOW statement accepts macro language keywords and does not attempt to resolve macro calls and macro variable references:

```
%window ask
    #5 @5 'Describe the problem.'
```

```
        #7 @5 'Enter /// when you are finished.'
        #9 @5 val 100 attr=underline;

%macro ask2;
      %global myprob;
      %local temp;

      %do %until(%superq(val) eq %str(///)); /* No ampersand */

          %display ask;
          %let temp=&temp %superq(val);    /* No ampersand */
      %end;
      %let myprob=&temp
%mend ask2;
```

You can invoke the macro ASK2 and enter a response:

```
%ask2

Try entering:
My macro ADDRESS starts with %MACRO ADDRESS(COMPANY,
CITY);. I called it with %ADDRESS(SMITH-JONES, INC., BOSTON),
but it said I had too many parameters. What happened?
///
```

The response contains a macro language keyword, a macro invocation, and unmatched parentheses.

Summary of Macro Quoting Functions and the Characters that They Mask

Different macro quoting functions mask different special characters and mnemonics so the macro facility interprets them as text instead of as macro language symbols.

The following table divides the symbols into categories and shows which macro quoting functions mask which symbols.

Table 7.6 *Summary of Special Characters and Macro Quoting Functions by Item*

<table>
<tr><th>Group</th><th>Items</th><th>Macro Quoting Functions</th></tr>
<tr><td>A</td><td>+ — */<>=¬^|~;, # blank AND OR NOT EQ NE LE LT GE GT IN</td><td>all</td></tr>
<tr><td>B</td><td>&%</td><td>%NRSTR, %NRBQUOTE, %SUPERQ, %NRQUOTE</td></tr>
<tr><td>C</td><td>unmatched' “()</td><td>%BQUOTE, %NRBQUOTE, %SUPERQ, %STR*, %NRSTR*, %QUOTE*, %NRQUOTE*</td></tr>
</table>

Table 7.7 By Function

Function	Affects Groups	Works At
%STR	A, C*	Macro compilation
%NRSTR	A, B, C*	Macro compilation
%BQUOTE	A, C	Macro execution
%NRBQUOTE	A, B, C	Macro execution
%SUPERQ	A, B, C	Macro execution (prevents resolution)
%QUOTE	A, C*	Macro execution. Requires unmatched quotation marks and parentheses to be marked with a percent sign (%).
%NRQUOTE	A, B, C*	Macro execution. Requires unmatched quotation marks and parentheses to be marked with a percent sign (%).

*Unmatched quotation marks and parentheses must be marked with a percent sign (%) when used with %STR, %NRSTR, %QUOTE, and %NRQUOTE.

Unquoting Text

Restoring the Significance of Symbols

To *unquote* a value means to restore the significance of symbols in an item that was previously masked by a macro quoting function.

Usually, after an item has been masked by a macro quoting function, it retains its special status until one of the following occurs:

- You enclose the item with the %UNQUOTE function. (See "%UNQUOTE Function" on page 274.)
- The item leaves the word scanner and is passed to the DATA step compiler, SAS procedures, SAS macro facility, or other parts of the SAS System.
- The item is returned as an unquoted result by the %SCAN, %SUBSTR, or %UPCASE function. (To retain a value's masked status during one of these operations, use the %QSCAN, %QSUBSTR, or %QUPCASE function. See "Other Functions That Perform Macro Quoting" on page 98 for more details.)

As a rule, you do not need to unquote an item because it is automatically unquoted when the item is passed from the word scanner to the rest of SAS. Under two circumstances, however, you might need to use the %UNQUOTE function to restore the original significance to a masked item:

- when you want to use a value with its restored meaning later in the same macro in which its value was previously masked by a macro quoting function
- when, as in a few cases, masking text with a macro quoting function changes the way the word scanner tokenizes it, producing SAS statements that look correct but that the SAS compiler does not recognize

Example of Unquoting

The following example illustrates using a value twice: once in macro quoted form and once in unquoted form. Suppose the macro ANALYZE is part of a system that enables you to compare the output of two statistical models interactively. First, you enter an operator to specify the relationship that you want to test (one result greater than another, equal to another, and so on). The macro ANALYZE tests the macro quoted value of the operator to verify that you have entered it correctly, uses the unquoted value to compare the values indicated, and writes a message. Match the numbers in the comments to the paragraphs below.

```
%macro analyze(stat);
   data _null_;
      set out1;
      call symput('v1',&stat);
   run;

   data _null_;
      set out2;
      call symput('v2',&stat);
   run;

   %put Preliminary test. Enter the operator.;
   %input;
   %let op=%bquote(&sysbuffr);
   %if &op=%str(=<) %then %let op=%str(<=);
   %else %if &op=%str(=>) %then %let op=%str(>=);
   %if &v1 %unquote(&op) &v2 %then
      %put You might proceed with the analysis.;
   %else
      %do;
         %put &stat from out1 is not &op &stat from out2.;
         %put Please check your previous models.;
      %end;
%mend analyze;
```

You mask the value of SYSBUFFR with the %BQUOTE function, which masks resolved items including unmatched, unmarked quotation marks and parentheses (but excluding the ampersand and percent sign).

The %IF condition compares the value of the macro variable OP to a string to see whether the value of OP contains the correct symbols for the operator. If the value contains symbols in the wrong order, the %THEN statement corrects the symbols. Because a value masked by a macro quoting function remains masked, you do not need to mask the reference &OP in the left side of the %IF condition.

Because you can see the characters in the right side of the %IF condition and in the %LET statement when you define the macro, you can use the %STR function to mask them. Masking them once at compilation is more efficient than masking them at each execution of ANALYZE.

To use the value of the macro variable OP as the operator in the %IF condition, you must restore the meaning of the operator with the %UNQUOTE function.

What to Do When Automatic Unquoting Does Not Work

When the macro processor generates text from an item masked by a macro quoting function, you can usually allow SAS to unquote the macro quoted items automatically. For example, suppose you define a macro variable PRINTIT:

```
%let printit=%str(proc print; run;);
```

Then you use that macro variable in your program:

```
%put *** This code prints the data set: &printit ***;
```

When the macro processor generates the text from the macro variable, the items masked by macro quoting functions are automatically unquoted, and the previously masked semicolons work normally when they are passed to the rest of SAS.

In rare cases, masking text with a macro quoting function changes the way the word scanner tokenizes the text. (The word scanner and tokenization are discussed in "SAS Programs and Macro Processing" on page 11 and "Macro Processing" on page 33.) For example, a single or double quotation mark produced by resolution within the %BQUOTE function becomes a separate token. The word scanner does not use it as the boundary of a literal token in the input stack. If generated text that was once masked by the %BQUOTE function looks correct but SAS does not accept it, you might need to use the %UNQUOTE function to restore normal tokenization.

How Macro Quoting Works

When the macro processor masks a text string, it masks special characters and mnemonics within the coding scheme, and prefixes and suffixes the string with a hexadecimal character, called a *delta character*. The prefix character marks the beginning of the string and also indicates what type of macro quoting is to be applied to the string. The suffix character marks the end of the string. The prefix and suffix characters preserve any leading and trailing blanks contained by the string. The hexadecimal characters used to mask special characters and mnemonics and the characters used for the prefix and suffix might vary and are not portable.

There are more hexadecimal combinations possible in each byte than are needed to represent the symbols on a keyboard. Therefore, when a macro quoting function recognizes an item to be masked, the macro processor uses a previously unused hexadecimal combination for the prefix and suffix characters.

Macro functions, such as %EVAL and %SUBSTR, ignore the prefix and suffix characters. Therefore, the prefix and suffix characters do not affect comparisons.

When the macro processor is finished with a macro quoted text string, it removes the macro quoting-coded substitute characters and replaces them with the original characters. The unmasked characters are passed on to the rest of the system. Sometimes you might see a message about unmasking, as in the following example:

```
/* Turn on SYMBOLGEN so you can see the messages about unquoting. */
options symbolgen;

/* Assign a value to EXAMPLE that contains several special */
/* characters and a mnemonic. */
```

```
%let example = %nrbquote( 1 + 1 = 3 Today's Test and More );

%put *&example*;
```

When this program is submitted, the following appears in the SAS log:

```
SYMBOLGEN:   Macro variable EXAMPLE resolves to  1 + 1 = 3 Today's
             Test and More
SYMBOLGEN:   Some characters in the above value which were subject
             to macro quoting have been unquoted for printing.
* 1 + 1 = 3 Today's Test and More *
```

As you can see, the leading and trailing blanks and special characters were retained in the variable's value. While the macro processor was working with the string, the string actually contained coded characters that were substituted for the "real" characters. The substitute characters included coded characters to represent the start and end of the string. The leading and trailing blanks were preserved. Characters were also substituted for the special characters `+`, `=`, and `'`, and the mnemonic **AND**. When the macro finished processing and the characters were passed to the rest of SAS, the coding was removed and the real characters were replaced.

"Unquoting Text " on page 95 provides more information about what happens when a masked string is unquoted. For more information, see "SYMBOLGEN System Option" on page 358.

Other Functions That Perform Macro Quoting

Functions That Start with the Letter Q

Some macro functions are available in pairs, where one function starts with the letter Q:

- %SCAN and %QSCAN
- %SUBSTR and %QSUBSTR
- %UPCASE and %QUPCASE
- %SYSFUNC and %QSYSFUNC

The Q*xxx* functions are necessary because by default, macro functions return an unquoted result, even if the argument was masked by a macro quoting function. The %QSCAN, %QSUBSTR, %QUPCASE, and %QSYSFUNC functions mask the returned value at execution time. The items masked are the same as the items masked by the %NRBQUOTE function.

Example Using the %QSCAN Function

The following macro uses the %QSCAN function to assign items in the value of SYSBUFFR (described in "Automatic Macro Variables" on page 192) as the values of separate macro variables. The numbers in the comments correspond to the explanations in the list that follows the macro code.

```
%macro splitit;
   %put What character separates the values?;  1
   %input;
   %let s=%bquote(&sysbuffr);  2
   %put Enter three values.;
```

```
   %input;
   %local i;
   %do i=1 %to 3;  3
      %global x&i;
      %let x&i=%qscan(%superq(sysbuffr),&i,&s);
4
   %end;
%mend splitit;

%splitit
What character separates the values?
#
Enter three values.
Fischer Books#Smith&Sons#Sarah's Sweet Shoppe
5
```

1. This question asks you to input a delimiter for the %QSCAN function that does not appear in the values that you enter.
2. Masking the value of SYSBUFFR with the %BQUOTE function enables you to choose a quotation mark or parenthesis as a delimiter if necessary.
3. The iterative %DO loop creates a global macro variable for each segment of SYSBUFFR and assigns it the value of that segment.
4. The %SUPERQ function masks the value of SYSBUFFR in the first argument of the %QSCAN function. It prevents any resolution of the value of SYSBUFFR.
5. The %QSCAN function returns macro quoted segments of the value of SYSBUFFR. Thus, the unmatched quotation mark in **`Sarah's Sweet Shoppe`** and the &*name* pattern in **`Smith&Sons`** do not cause problems.

Chapter 8
Interfaces with the Macro Facility

Interfaces with the Macro Facility

An *interface* with the macro facility is not part of the macro processor but rather a SAS software feature that enables another portion of the SAS language to interact with the macro facility during execution. For example, a DATA step interface enables you to access macro variables from the DATA step. Macro facility interfaces are useful because, in general, macro processing happens before DATA step, SQL, SCL, or SAS/CONNECT execution, so the connection between the macro facility and the rest of SAS is not usually dynamic. But by using an interface to the macro facility, you can dynamically connect the macro facility to the rest of SAS.

Note: The %SYSFUNC and %QSYSFUNC macro functions enable you to use SAS language functions with the macro processor. The %SYSCALL macro statement enables you to use SAS language CALL routines with the macro processor. While these elements of the macro language are not considered true macro facility interfaces, they are discussed in this section. See "Macro Language Elements" on page 155 for more information about these macro language elements.

DATA Step Interfaces

Interacting With the Macro Facility During DATA Step Execution

DATA step interfaces consist of eight tools that enable a program to interact with the macro facility during DATA step execution. Because the work of the macro facility takes place before DATA step execution begins, information provided by macro statements has already been processed during DATA step execution. You can use one of the DATA step interfaces to interact with the macro facility during DATA step execution. You can use DATA step interfaces to do the following:

- pass information from a DATA step to a subsequent step in a SAS program
- invoke a macro based on information available only when the DATA step executes
- resolve a macro variable while a DATA step executes
- delete a macro variable
- pass information about a macro variable from the macro facility to the DATA step

The following table lists the DATA step interfaces by category and their uses.

Table 8.1 *DATA Step Interfaces to the Macro Facility*

Category	Tool	Description
Execution	CALL EXECUTE routine	Resolves its argument and executes the resolved value at the next step boundary (if the value is a SAS statement) or immediately (if the value is a macro language element).
Resolution	RESOLVE function	Resolves the value of a text expression during DATA step execution.
Deletion	CALL SYMDEL routine	Deletes the indicated macro variable named in the argument.
Information	SYMEXIST function	Returns an indication as to whether the macro variable exists.
Read or Write	SYMGET function	Rreturns the value of a macro variable during DATA step execution.
Information	SYMGLOBL function	Returns an indication as to whether the macro variable is global in scope.
Information	SYMLOCAL function	Returns an indication as to whether the macro variable is local in scope.

Category	Tool	Description
Read or Write	CALL SYMPUT routine	Assigns a value produced in a DATA step to a macro variable.

CALL EXECUTE Routine Timing Details

CALL EXECUTE is useful when you want to execute a macro conditionally. But you must remember that if CALL EXECUTE produces macro language elements, those elements execute immediately. If CALL EXECUTE produces SAS language statements, or if the macro language elements generate SAS language statements, those statements execute after the end of the DATA step's execution.

Note: Because macro references execute immediately and SAS statements do not execute until after a step boundary, you cannot use CALL EXECUTE to invoke a macro that contains references for macro variables that are created by CALL SYMPUT in that macro.

Example of Using CALL EXECUTE Incorrectly

In this example, the CALL EXECUTE routine is used incorrectly:

```
data prices;  /* ID for price category and actual price */
   input code amount;
   datalines;
56 300
99 10000
24 225
;

%macro items;
   %global special;
   %let special=football;
%mend items;

data sales;    /* incorrect usage */
   set prices;
   length saleitem $ 20;
   call execute('%items');
   saleitem="&special";
run;
```

In the DATA SALES step, the assignment statement for SALEITEM requires the value of the macro variable SPECIAL at DATA step compilation. CALL EXECUTE does not produce the value until DATA step execution. Thus, you receive a message about an unresolved macro variable, and the value assigned to SALEITEM is **`&special`**.

In this example, it would be better to eliminate the macro definition (the %LET macro statement is valid in open code) or move the DATA SALES step into the macro ITEMS. In either case, CALL EXECUTE is not necessary or useful. Here is one version of this program that works:

```
data prices;   /* ID for price category and actual price */
   input code amount;
   datalines;
```

```
56 300
99 10000
24 225
;

%let special=football;  /* correct usage */

data sales;
   set prices;
   length saleitem $ 20;
   saleitem="&special";
run;
```

The %GLOBAL statement is not necessary in this version. Because the %LET statement is executed in open code, it automatically creates a global macro variable. (See "Scopes of Macro Variables" on page 43 for more information about macro variable scopes.)

Example of Common Problem with CALL EXECUTE

This example shows a common pattern that causes an error.

```
/* This version of the example shows the problem. */

data prices;          /* ID for price category and actual price */
   input code amount;
   cards;
56 300
99 10000
24 225
;
data names;      /* name of sales department and item sold */
   input dept $ item $;
   datalines;
BB  Boat
SK  Skates
;

%macro items(codevar=);  /* create macro variable if needed */
   %global special;
   data _null_;
      set names;
      if &codevar=99 and dept='BB' then call symput('special', item);
   run;
%mend items;

data sales;  /* attempt to reference macro variable fails */
   set prices;
   length saleitem $ 20;
   if amount > 500 then
      call execute('%items(codevar=' || code || ')' );
   saleitem="&special";
run;
```

In this example, the DATA SALES step still requires the value of SPECIAL during compilation. The CALL EXECUTE routine is useful in this example because of the conditional IF statement. But as in the first example, CALL EXECUTE still invokes the macro ITEMS during DATA step execution — not during compilation. The macro

ITEMS generates a DATA _NULL_ step that executes after the DATA SALES step has ceased execution. The DATA _NULL_ step creates SPECIAL, and the value of SPECIAL is available after the _NULL_ step ceases execution, which is much later than when the value was needed.

This version of the example corrects the problem:

```
/* This version solves the problem. */

data prices;      /* ID for price category and actual price */
   input code amount;
   datalines;
56 300
99 10000
24 225
;

data names;      /* name of sales department and item sold */
   input dept $ item $;
   cards;
BB  Boat
SK  Ski
;
%macro items(codevar=);    /* create macro variable if needed */
   %global special;
   data _null_;
      set names;
      if &codevar=99 and dept='BB' then
         call symput('special', item);
   run;
%mend items;

data _null_;     /* call the macro in this step */
   set prices;
   if amount > 500 then
      call execute('%items(codevar=' || code || ')' );
run;

data sales;     /* use the value created by the macro in this step */
   set prices;
   length saleitem $ 20;
   saleitem="&special";
run;
```

This version uses one DATA _NULL_ step to call the macro ITEMS. After that step ceases execution, the DATA _NULL_ step generated by ITEMS executes and creates the macro variable SPECIAL. Then the DATA SALES step references the value of SPECIAL as usual.

Using SAS Language Functions in the DATA Step and Macro Facility

The macro functions %SYSFUNC and %QSYSFUNC can call SAS language functions and functions written with SAS/TOOLKIT software to generate text in the macro facility. %SYSFUNC and %QSYSFUNC have one difference: the %QSYSFUNC masks special characters and mnemonics and %SYSFUNC does not. For more information about these functions, see "%SYSFUNC and %QSYSFUNC Functions" on page 268.

%SYSFUNC arguments are a single SAS language function and an optional format. See the following examples:

```
%sysfunc(date(),worddate.)
%sysfunc(attrn(&dsid,NOBS))
```

You cannot nest SAS language functions within %SYSFUNC. However, you can nest %SYSFUNC functions that call SAS language functions, as in the following statement:

```
%sysfunc(compress(%sysfunc(getoption(sasautos)),%str(%)%(%')))
```

This example returns the value of the SASAUTOS= system option, using the COMPRESS function to eliminate opening parentheses, closing parentheses, and single quotation marks from the result. Note the use of the %STR function and the unmatched parentheses and quotation marks that are marked with a percent sign (%).

All arguments in SAS language functions within %SYSFUNC must be separated by commas. You cannot use argument lists preceded by the word OF.

Because %SYSFUNC is a macro function, you do not need to enclose character values in quotation marks as you do in SAS language functions. For example, the arguments to the OPEN function are enclosed in quotation marks when the function is used alone but do not require quotation marks when used within %SYSFUNC.

Here are some examples of the contrast between using a function alone and within %SYSFUNC:

- **`dsid = open("sasuser.houses","i");`**
- **`dsid = open("&mydata","&mode");`**
- **`%let dsid = %sysfunc(open(sasuser.houses,i));`**
- **`%let dsid = %sysfunc(open(&mydata,&mode));`**

You can use %SYSFUNC and %QSYSFUNC to call all of the DATA step SAS functions except the ones that are listed in table Table 17.2 on page 269. In the macro facility, SAS language functions called by %SYSFUNC can return values with a length up to 32K. However, within the DATA step, return values are limited to the length of a data set character variable.

The %SYSCALL macro statement enables you to use SAS language CALL routines with the macro processor, and it is described in "Macro Statements" on page 281.

Interfaces with the SQL Procedure

Using PROC SQL

Structured Query Language (SQL) is a standardized, widely used language for retrieving and updating data in databases and relational tables. SAS software's SQL processor enables you to do the following:

- create tables and views
- retrieve data stored in tables
- retrieve data stored in SQL and SAS/ACCESS views

- add or modify values in tables
- add or modify values in SQL and SAS/ACCESS views

INTO Clause

SQL provides the INTO clause in the SELECT statement for creating SAS macro variables. You can create multiple macro variables with a single INTO clause. The INTO clause follows the same scoping rules as the %LET statement. See “Macro Variables” on page 19 for a summary of how macro variables are created. For further details and examples relating to the INTO clause, see “INTO Clause” on page 277.

Controlling Job Execution

PROC SQL also provides macro tools to do the following:

- stop execution of a job if an error occurs
- execute programs conditionally based on data values

The following table provides information about macro variables created by SQL that affect job execution.

Table 8.2 *Macro Variables that Affect Job Execution*

Macro Variable	Description
SQLEXITCODE	Contains the highest return code that occurred from some types of SQL insert failures. This return code is written to the SYSERR macro variable when PROC SQL terminates.
SQLOBS	Contains the number of rows or observations produced by a SELECT statement.
SQLOOPS	Contains the number of iterations that the inner loop of PROC SQL processes.
SQLRC	Contains the return code from an SQL statement. For return codes, see SAS SQL documentation.
SQLXMSG	Contains descriptive information and the DBMS-specific return code for the error that is returned by the pass-through facility.
SQLXRC	contains the DBMS-specific return code that is returned by the pass-through facility.

Interfaces with the SAS Component Language

Using an SCL Program

You can use the SAS macro facility to define macros and macro variables for an SCL program. Then, you can pass parameters between macros and the rest of the program.

Also, through the use of the autocall and compiled stored macro facilities, macros can be used by more than one SCL program.

Note: Macro modules can be more complicated to maintain than a program segment because of the symbols and macro quoting that might be required. Also, implementing modules as macros does not reduce the size of the compiled SCL code. Program statements generated by a macro are added to the compiled code as if those lines existed at that location in the program.

The following table lists the SCL macro facility interfaces.

Table 8.3 *SCL Interfaces to the Macro Facility*

Category	Tool	Description
Read or Write	SYMGET	returns the value of a global macro variable during SCL execution.
	SYMGETN	returns the value of a global macro variable as a numeric value.
	CALL SYMPUT	assigns a value produced in SCL to a global macro variable.
	CALL SYMPUTN	assigns a numeric value to a global macro variable.

Note: It is inefficient to use SYMGETN to retrieve values that are not assigned with SYMPUTN. It is also inefficient to use `&` to reference a macro variable that was created with CALL SYMPUTN. Instead, use SYMGETN. In addition, it is inefficient to use SYMGETN and CALL SYMPUTN with values that are not numeric.

For details about these elements, see "DATA Step Call Routines for Macros" on page 221 and "DATA Step Functions for Macros" on page 231.

How Macro References Are Resolved by SCL

An important point to remember when using the macro facility with SCL is that macros and macro variable references in SCL programs are resolved when the SCL program compiles, not when you execute the application. To further control the assignment and resolution of macros and macro variables, use the following techniques:

- If you want macro variables to be assigned and retrieved when the SCL program executes, use CALL SYMPUT and CALL SYMPUTN in the SCL program.
- If you want a macro call or macro variable reference to resolve when an SCL program executes, use SYMGET and SYMGETN in the SCL program.

Referencing Macro Variables in Submit Blocks

In SCL, macro variable references are resolved at compile time unless they are in a Submit block. When SCL encounters a name prefixed with an ampersand (&) in a Submit block, it checks whether the name following the ampersand is the name of an

SCL variable. If so, SCL substitutes the value of the corresponding variable for the variable reference in the submit block. If the name following the ampersand does not match any SCL variable, the name passes intact (including the ampersand) with the submitted statements. When SAS processes the statements, it attempts to resolve the name as a macro variable reference.

To guarantee that a name is passed as a macro variable reference in submitted statements, precede the name with two ampersands (for example, &&DSNAME). If you have both a macro variable and an SCL variable with the same name, a reference with a single ampersand substitutes the SCL variable. To force the macro variable to be substituted, reference it with two ampersands (&&).

Considerations for Sharing Macros between SCL Programs

Sharing macros between SCL programs can be useful, but it can also raise some configuration management problems. If a macro is used by more than one program, you must keep track of all the programs that use it so you can recompile all of them each time the macro is updated. Because SCL is compiled, each SCL program that calls a macro must be recompiled whenever that macro is updated.

CAUTION:

Recompile the SCL program. If you fail to recompile the SCL program when you update the macro, you run the risk of the compiled SCL being out of sync with the source.

Example Using Macros in an SCL Program

This SCL program is for an example application with the fields BORROWED, INTEREST, and PAYMENT. The program uses the macros CKAMOUNT and CKRATE to validate values entered into fields by users. The program calculates the payment, using values entered for the interest rate (INTEREST) and the sum of money (BORROWED).

```
/* Display an error message if AMOUNT */
   /* is less than zero or larger than 1000. */
%macro ckamount(amount);
   if (&amount < 0) or (&amount > 1000) then
      do;
         erroron borrowed;
         _msg_='Amount must be between $0 and $1,000.';
         stop;
      end;
   else erroroff borrowed;
%mend ckamount;

   /* Display an error message if RATE */
   /* is less than 0 or greater than 1.5 */
%macro ckrate(rate);
   if (&rate < 0) or (&rate > 1) then
      do;
         erroron interest;
         _msg_='Rate must be between 0 and 1.5';
         stop;
      end;
   else erroroff interest;
%mend ckrate;
```

```
   /*  Open the window with BORROWED at 0 and INTEREST at .5.  */
INIT:
   control error;
   borrowed=0;
   interest=.5;
return;

MAIN:
      /*  Run the macro CKAMOUNT to validate  */
      /*  the value of BORROWED.              */
   %ckamount(borrowed)
      /*  Run the macro CKRATE to validate  */
      /* the value of INTEREST.             */
   %ckrate(interest)
      /*  Calculate payment.  */
   payment=borrowed*interest;
return;

TERM:
return;
```

SAS/CONNECT Interfaces

Using %SYSRPUT with SAS/CONNECT

The %SYSRPUT macro statement is submitted with SAS/CONNECT to a remote host to retrieve the value of a macro variable stored on the remote host. %SYSRPUT assigns that value to a macro variable on the local host. %SYSRPUT is similar to the %LET macro statement because it assigns a value to a macro variable. However, %SYSRPUT assigns a value to a variable on the local host, not on the remote host where the statement is processed. The %SYSRPUT statement places the macro variable in the current scope of the local host.

Note: The names of the macro variables on the remote and local hosts must not contain a leading ampersand.

The %SYSRPUT statement is useful for capturing the value of the automatic macro variable SYSINFO and passing that value to the local host. SYSINFO contains return-code information provided by some SAS procedures. Both the UPLOAD and the DOWNLOAD procedures of SAS/CONNECT can update the macro variable SYSINFO and set it to a nonzero value when the procedure terminates due to errors. You can use %SYSRPUT on the remote host to send the value of the SYSINFO macro variable back to the local SAS session. Thus, you can submit a job to the remote host and test whether a PROC UPLOAD or DOWNLOAD step has successfully completed before beginning another step on either the remote host or the local host.

To use %SYSRPUT, you must have invoked a remote SAS windowing environment session by submitting the DMR option with the SAS command. For details about using %SYSRPUT, see the SAS/CONNECT documentation.

To create a new macro variable or to modify the value of an existing macro variable on a remote host or a server, use the %SYSLPUT macro statement.

Example Using %SYSRPUT to Check the Value of a Return Code on a Remote Host

This example illustrates how to download a file and return information about the success of the step. When remote processing is completed, the job checks the value of the return code stored in RETCODE. Processing continues on the local host if the remote processing is successful. In this example, the %SYSRPUT statement follows a PROC DOWNLOAD step, so the value returned by SYSINFO indicates the success of the PROC DOWNLOAD step:

```
/* This code executes on the remote host. */
rsubmit;
   proc download data=remote.mydata out=local.mydata;
   run;
         /* RETCODE is on the local host. */
         /* SYSINFO is on the remote host. */
   %sysrput retcode=&sysinfo;
endrsubmit;

   /* This code executes on the local host. */
%macro checkit;
   %if &retcode = 0 %then
      %do;
         further processing on local host
      %end;
%mend checkit;

%checkit
```

To determine the success or failure of a step executed on a remote host, use the %SYSRPUT macro statement to check the value of the automatic macro variable SYSERR.

For more details and syntax of the %SYSRPUT statement, see "%SYSRPUT Statement" on page 319.

Using %SYSLPUT with SAS/CONNECT

The %SYSLPUT statement is a macro statement that is submitted in the client session to assign a value that is available in the client session to a macro variable that can be accessed from the server session. If you are signed on to multiple server sessions, %SYSLPUT submits the macro assignment statement to the most recently used server session. If you are signed on to only one server session, %SYSLPUT submits the macro assignment statement to that server session. If you are not signed on to any session, an error condition results. Like the %LET statement, the %SYSLPUT statement assigns a value to a macro variable. Unlike %LET, the %SYSRPUT statement assigns a value to a variable in the server session rather than in the client session where the statement is executed. The %SYSRPUT statement stores the macro variable in the Global Symbol Table in the server session.

For details about using %SYSLPUT, see the SAS/CONNECT documentation.

Example Using %SYSLPUT

%SYSLPUT enables you to dynamically assign values to variables that are used by macros that are executed in a server session. The macro statement %SYSLPUT is used to create the macro variable REMID in the server session and to use the value of the client macro variable RUNID. The REMID variable is used by the %DOLIB macro, which is executed in a server session, to find out which operating system-specific library assignment should be used in the server session.

```
%macro assignlib (runid);

   signon rem &runid
   %syslput remid=&runid
   rsubmit rem &runid
      %macro dolib;
         %if (&runid eq 1) %then %do;
            libname mylib 'h:';
            %end;
         %else %if (&runid eq 2) %then %do;
            libname mylib '/afs/some/unix/path';
            %end;
      %mend;
      %dolib;
   endrsubmit;

%mend;
```

Chapter 9
Storing and Reusing Macros

Storing and Reusing Macros

When you submit a macro definition, by default, the macro processor compiles and stores the macro in a SAS catalog in the WORK library. These macros, referred to as *session compiled macros*, exist only during the current SAS session. To save frequently used macros between sessions, you can use either the autocall macro facility or the stored compiled macro facility.

The autocall macro facility stores the source for SAS macros in a collection of external files called an *autocall library*. The autocall facility is useful when you want to create a pool of easily maintained macros in a location that can be accessed by different applications and users. Autocall libraries can be concatenated together. The primary disadvantage of the autocall facility is that the first time an autocall macro is called in a session, the macro processor compiles it. This compilation is overhead that you can avoid by using the stored compiled macro facility.

The stored compiled macro facility stores compiled macros in a SAS catalog in a SAS library that you specify. By using stored compiled macros, you might save macro compilation time in your production-level jobs. However, because these stored macros are compiled, you must save and maintain the source for the macro definitions in a different location.

The autocall facility and the stored compiled macro facility each offer advantages. Here are some of the factors that determine how you choose to save a macro definition:

- how often you use a macro
- how often you change it

- how many users need to execute it
- how many compiled macro statements it has

If you are developing new programs, consider creating macros and compiling them during your current session. If you are running production-level jobs using name-style macros, consider using stored compiled macros. If you are letting a group of users share macros, consider using the autocall facility.

Note: For greater efficiency, store only name-style macros if you use the stored compiled macro facility. Storing statement-style and command-style macros is less efficient.

It is good practice, when you are programming stored compiled macros or autocall macros, to use the %LOCAL statement to define macro variables that will be used only inside that macro. Otherwise, values of macro variables defined outside of the current macro might be altered. See the discussion of macro variable scopes in "Scopes of Macro Variables" on page 43.

In general, macro and variable names in the SAS macro facility are case insensitive and are internally changed to uppercase. The values are case sensitive in the SAS macro facility and are not changed.

When calling an autocall macro or a stored compiled macro, the macro name is changed to uppercase and passed to the catalog routines to open a member of that name. The catalog routines are host dependent and use the default casing for the particular host when searching for a member. Macro catalog entries should be made using the default casing for the host in question. Here are the host defaults:

- UNIX default is lowercase
- z/OS default is uppercase
- Windows default is lowercase

Note: In UNIX, the member name that contains the autocall macro must be all lowercase letters.

Saving Macros in an Autocall Library

Overview of an Autocall Library

Generally, an autocall library is a directory containing individual files, each of which contains one macro definition. In SAS 6.11 and later, an autocall library can also be a SAS catalog. (See the following section for more information about using SAS catalogs as autocall libraries.)

Operating Environment Information
Autocall Libraries on Different Hosts The term *directory* refers to an aggregate storage location that contains files (or members) managed by the host operating system. Different host operating systems identify an aggregate storage location with different names, such as a directory, a subdirectory, a maclib, a text library, or a partitioned data set. For more information, see the SAS Companion for your operating system.

Using Directories as Autocall Libraries

To use a directory as a SAS autocall library, do the following:

1. To create library members, store the source code for each macro in a separate file in a directory. The name of the file must be the same as the macro name. For example, the statements defining a macro that you would call by submitting %SPLIT must be in a file named SPLIT.

   Operating Environment Information
   Autocall Library Member Names On operating systems that allow filenames with extensions, you must name autocall macro library members with a special extension, usually **.SAS**. Look at the autocall macros on your system provided by SAS to determine whether names of files containing macros must have a special extension at your site. On z/OS operating systems, you must assign the macro name as the name of the PDS member.

2. Set the SASAUTOS system option to specify the directory as an autocall library. On most hosts, the reserved fileref SASAUTOS is assigned at invocation time to the autocall library supplied by SAS or another one designated by your site. If you are specifying one or more autocall libraries, remember to concatenate the autocall library supplied by SAS with your autocall libraries so that these macros will also be available. For details, refer to your host documentation and "SASAUTOS= System Option" on page 356.

When storing files in an autocall library, remember the following:

- Although SAS does not restrict the type of material that you place in an autocall library, you should store only autocall library files in it to avoid confusion and for ease of maintenance.
- Although SAS lets you include more than one macro definition, as well as open code, in an autocall library member, you should generally keep only one macro in any autocall library member. If you need to keep several macros in the same autocall library member, keep related macros together.

Using SAS Catalogs as Autocall Libraries

In SAS 6.11 and later, you can use the CATALOG access method to store autocall macros as SOURCE entries in SAS catalogs. To create an autocall library using a SAS catalog, follow these steps:

1. Use a LIBNAME statement to assign a libref to the SAS library.
2. Use a FILENAME statement with the CATALOG argument to assign a fileref to the catalog that contains the autocall macros. For example, the following code creates a fileref, MYMACROS, that points to a catalog named MYMACS.MYAUTOS:

   ```
   libname mymacs 'SAS-data-library';
   filename mymacros catalog 'mymacs.myautos';
   ```

3. Store the source code for each macro in a SOURCE entry in a SAS catalog. (SOURCE is the entry type.) The name of the SOURCE entry must be the same as the macro name.
4. Set the SASAUTOS system option to specify the fileref as an autocall library. For more information, see "SASAUTOS= System Option" on page 356.

Calling an Autocall Macro

To call an autocall macro, the system options MAUTOSOURCE must be set and SASAUTOS must be assigned. MAUTOSOURCE enables the autocall facility, and SASAUTOS specifies the autocall libraries. For more information, see "MAUTOSOURCE System Option" on page 335 and "SASAUTOS= System Option" on page 356.

Once you have set the required options, calling an autocall macro is like calling a macro that you have created in your current session. However, it is important that you understand how the macro processor locates the called macro. When you call a macro, the macro processor does the following tasks:

- searches for a session compiled macro definition
- searches for a stored compiled macro definition in the library specified by the SASMSTORE option, if the MSTORED option is set
- searches for a member in the autocall libraries specified by the SASAUTOS option in the order in which they are specified, if the MAUTOSOURCE option is set
- searches the SASHELP library for SAS production stored compiled macro definitions

When SAS finds a library member in an autocall library with that macro name, the macro processor does the following:

- compiles all of the source statements in that member, including any and all macro definitions, and stores the result in the session catalog
- executes any open code (macro statements or SAS source statements not within any macro definition) in that member
- executes the macro with the name that you invoked

Note: If an autocall library member contains more than one macro, the macro processor compiles all of the macros but executes only the macro with the name that you invoked.

Any open code statements in the same autocall library member as a macro execute only the first time you invoke the macro. When you invoke the macro later in the same session, the compiled macro is executed, which contains only the compiled macro definition and not the other code the autocall macro source file might have contained.

It is not advisable to change SASAUTOS during a SAS session. If you change the SASAUTOS= specification in an ongoing SAS session, SAS will store the new specification only until you invoke an uncompiled autocall macro and then will close all opened libraries and open all the newly specified libraries that it can open.

For information about debugging autocall macros, see "Macro Facility Error Messages and Debugging" on page 120.

Saving Macros Using the Stored Compiled Macro Facility

Overview of the Stored Compiled Macro Facility

The stored compiled macro facility compiles and saves compiled macros in a permanent catalog in a library that you specify. This compilation occurs only once. If the stored compiled macro is called in the current or later sessions, the macro processor executes the compiled code.

In SAS 9.1.3 or higher, the stored compiled macro catalog is initially opened for read-only access. When a stored compiled macro is being compiled or updated, the catalog is immediately closed and reopened for update access. After the macro is compiled and the catalog has been updated or changed, the catalog is again immediately closed and reopened for read-only access.

Compiling and Storing a Macro Definition

To compile a macro definition in a permanent catalog, you must first create the source for each stored compiled macro. To store the compiled macro, use the following steps:

1. Use the STORE option in the %MACRO statement. You can use the SOURCE option to store the source code with the compiled code. In addition, you can assign a descriptive title for the macro entry in the SAS catalog, by specifying the DES= option. For example, the %MACRO statement in the following definition shows the STORE, SOURCE, and DES= options:

   ```
   %macro myfiles / store source
        des='Define filenames';
     filename file1 'external-file-1';
     filename file2 'external-file-2';
   %mend;
   ```

   CAUTION:

   Save your macro source code. You cannot recreate the source statements from a compiled macro. Therefore, you must save the original macro source statements if you want to change the macro. For all stored compiled macros, you should document your macro source code well. You can save the source code with the compiled code using the SOURCE option in the %MACRO statement or you can save the source in a separate file. If you save the source in a separate file, it is recommended that you save the source code in the same catalog as the compiled macro. In this example, save it to the following library:

   ```
   mylib.sasmacro.myfiles.source
   ```

   Note: To retrieve the source of a compiled stored macro, see "%COPY Statement" on page 286.

2. Set the MSTORED system option to enable the stored compiled macro facility. For more information, see "MSTORED System Option" on page 354.

3. Assign the SASMSTORE option to specify the SAS library that contains or will contain the catalog of stored compiled SAS macros. For example, to store or call

compiled macros in a SAS catalog named MYLIB.SASMACR, submit these statements:

```
libname mylib 'SAS-data-library';
options mstored sasmstore=mylib;
```

For more information, see “SASMSTORE= System Option” on page 357.

4. Submit the source for each macro that you want to compile and permanently store.

You cannot move a stored compiled macro to another operating system or to a different release of SAS. You can, however, move the macro source code to another operating system or to a different SAS release where you can then compile and store it. For more information, see your host companion.

Storing Autocall Macros Supplied by SAS

If you use the macros in the autocall library supplied by SAS, you can save macro compile time by compiling and storing those macros in addition to ones that you create yourself. Many of the macros related to Base SAS software that are in the autocall library supplied by SAS can be compiled and stored in a SAS catalog named SASMACR by using the autocall macro COMPSTOR that is supplied by SAS. For more information, see “%COMPSTOR Autocall Macro” on page 177.

Calling a Stored Compiled Macro

Once you have set the required system options, calling a stored compiled macro is just like calling session compiled macros. However, it is important that you understand how the macro processor locates a macro. When you call a macro, the macro processor searches for the macro name using this sequence:

1. the macros compiled during the current session
2. the stored compiled macros in the SASMACR catalog in the specified library (if options MSTORED and SASMSTORE= are in effect)
3. each autocall library specified in the SASAUTOS option (if options SASAUTOS= and MAUTOSOURCE are in effect)
4. the stored compiled macros in the SASMACR catalog in the SASHELP library

You can display the entries in a catalog containing compiled macros. For more information, see “Macro Facility Error Messages and Debugging” on page 120.

Chapter 10

Macro Facility Error Messages and Debugging

General Macro Debugging Information

Developing Macros in a Layered Approach

Because the macro facility is such a powerful tool, it is also complex, and debugging large macro applications can be extremely time-consuming and frustrating. Therefore, it makes sense to develop your macro application in a way that minimizes the errors. This makes the errors that do occur as easy as possible to find and fix. The first step is to understand what type of errors can occur and when they manifest themselves. Then, develop your macros using a modular, layered approach. Finally, use some built-in tools such as system options, automatic macro variables, and the %PUT statement to diagnose errors.

Note: To receive certain important warning messages about unresolved macro names and macro variables, be sure the “SERROR System Option” on page 357 and “MERROR System Option” on page 340 are in effect. See “System Options for Macros” on page 329 for more information about these system options.

Encountering Errors

When the word scanner processes a program and finds a token in the form of `&` or `%`, it triggers the macro processor to examine the name token that follows the `&` or `%`. Depending on the token, the macro processor initiates one of the following activities:

- macro variable resolution
- macro open code processing
- macro compilation
- macro execution

An error can occur during any one of these stages. For example, if you misspell a macro function name or omit a necessary semicolon, that is a *syntax error* during compilation. Syntax errors occur when program statements do not conform to the rules of the macro language. Or, you might refer to a variable out of scope, causing a *macro variable resolution error*. *Execution errors* (also called *semantic errors*) are usually errors in program logic. They can occur, for example, when the text generated by the macro has faulty logic (statements not executed in the right order or in the way you expect).

Of course, even if your macro code is perfect, that does not guarantee that you will not encounter errors caused by plain SAS code. For example, you might encounter the following:

- a libref that is not defined
- a syntax error in open code (that is, outside of a macro definition)
- a typo in the code that your macro generates

Typically, error messages with numbers are plain SAS code error messages. Error messages generated by the macro processor do not have numbers.

Developing Bug-free Macros

When programming in any language, it is good technique to develop your code in modules. That is, instead of writing one massive program, develop it piece by piece, test each piece separately, and put the pieces together. This technique is especially useful when developing macro applications because of the two-part nature of SAS macros: macro code and the SAS code generated by the macro code.

Another good idea is to proofread your macro code for common mistakes before you submit it.

The following list outlines some key items to check for:

- The names in the %MACRO and %MEND statements match, and there is a %MEND for each %MACRO.
- The number of %DO statements matches the number of %END statements.
- %TO values for iterative %DO statements exist and are appropriate.
- All statements end with semicolons.
- Comments begin and end correctly and do not contain unmatched single quotation marks.
- Macro variable references begin with `&` and macro statements begin with `%`.
- Macro variables created by CALL SYMPUT are not referenced in the same DATA step in which they are created.
- Statements that execute immediately (such as %LET) are not part of conditional DATA step logic.
- Single quotation marks are not used around macro variable references (such as in TITLE or FILENAME statements). When used in quoted strings, macro variable references resolve only in strings marked with double quotation marks.
- Macro variable values do not contain any keywords or characters that could be interpreted as mathematical operators. (If they do contain such characters, use the appropriate macro quoting function.)
- Macro variables, %GOTO labels, and macro names do not conflict with reserved SAS and host environment keywords.

Troubleshooting Your Macros

Solving Common Macro Problems

The following table lists some problems that you might encounter when working with the macro facility. Because many of these problems do not cause error messages to be written to the SAS log, solving them can be difficult. For each problem, the table gives some possible causes and solutions.

Table 10.1 Commonly Encountered Macro Problems

Problem	Cause	Explanation
SAS windowing environment session stops responding after you submit a macro definition. You type and submit code but nothing happens.	• Syntax error in %MEND statement • Missing semicolon, parenthesis, or quotation mark • Missing %MEND statement • Unclosed comment	The %MEND statement is not recognized and all text is becoming part of the macro definition.
SAS windowing environment session stops responding after you call a macro.	An error in invocation, such as forgetting to provide one or more parameters, or forgetting to use parentheses when invoking a macro that is defined with parameters.	The macro facility is waiting for you to finish the invocation.
The macro does not compile when you submit it.	A syntax error exists somewhere in the macro definition.	Only syntactically correct macros are compiled.
The macro does not execute when you call it or partially executes and stops.	• A bad value was passed to the macro (for example, as a parameter). • A syntax error exists somewhere in the macro definition.	A macro successfully executes only when it receives the correct number of parameters that are of the correct type.
The macro executes but the SAS code gives bad results or no results.	Incorrect logic in the macro or SAS code.	
Code runs fine if submitted as open code, but when generated by a macro, the code does not work and issues strange error messages.	• Tokenization is not as you intended. • A syntax error exists somewhere in the macro definition.	Rarely, macro quoting functions alter the tokenization of text enclosed in them. Use the "%UNQUOTE Function" on page 274.
A %MACRO statement generates "invalid statement" error.	• The MACRO system option is turned off. • A syntax error exists somewhere in the macro definition.	For the macro facility to work, the MACRO system option must be on. Edit your SAS configuration file accordingly.

The following table lists some common macro error and warning messages. For each message, some probable causes are listed, and pointers to more information are provided.

Table 10.2 *Common Macro Error Messages and Causes*

Error Message	Possible Causes	For More Information
`Apparent invocation of macro xxx not resolved.`	• You have misspelled the macro name. • MAUTOSOURCE system option is turned off. • MAUTOSOURCE is on, but you have specified an incorrect pathname in the SASAUTOS= system option. • You are using the autocall facility but have given the macro and file different names. • You are using the autocall facility but did not give the file the `.sas` extension. • There is a syntax error within the macro definition.	• Check the spelling of the macro name. • "Solving Problems with the Autocall Facility" on page 130. • "Developing Bug-free Macros" on page 121.
`Apparent symbolic reference xxx not resolved.`	• You are trying to resolve a macro variable in the same DATA step as the CALL SYMPUT that created it. • You have misspelled the macro variable name. • You are referencing a macro variable that is not in scope. • You have omitted the period delimiter when adding text to the end of the macro variable.	• "Resolving Timing Issues" on page 128. • Check the spelling of the macro variable. • "Solving Problems with Macro Variable Scope" on page 124. • "Solving Macro Variable Resolution Problems" on page 123. • "Generating a Suffix for a Macro Variable Reference" in "Introduction to the Macro Facility" on page 3.

Solving Macro Variable Resolution Problems

When the macro processor examines a name token that follows an `&`, it searches the macro symbol tables for a matching macro variable entry. If it finds a matching entry, it pulls the associated text from the symbol table and replaces `&name` on the input stack. A macro variable name is passed to the macro processor, but the processor does not find a matching entry in the symbol tables. So, it leaves the token on the input stack and generates this message:

```
WARNING: Apparent symbolic reference
NAME not resolved.
```

The unresolved token is transferred to the input stack for use by other parts of SAS.

Note: You receive the WARNING only if the SERROR system option is on.

To solve these problems, check that you have spelled the macro variable name right and that you are referencing it in an appropriate scope.

When a macro variable resolves but does not resolve to the correct value, you can check several things. First, if the variable is a result of a calculation, ensure that the correct values were passed into the calculation. And, ensure that you have not inadvertently changed the value of a global variable. (See "Solving Problems with Macro Variable Scope" on page 124 for more details about variable scope problems.)

Another common problem is adding text to the end of a macro variable but forgetting to add a delimiter that shows where the macro variable name ends and the added text begins. For example, suppose you want to write a TITLE statement with a reference to WEEK1, WEEK2, and so on. You set a macro variable equal to the first part of the string and supply the week's number in the TITLE statement:

```
%let wk=week;

title "This is data for &wk1";    /* INCORRECT */
```

When these statements compile, the macro processor looks for a macro variable named WK1, not WK. To fix the problem, add a period (the macro delimiter) between the end of the macro variable name and the added text, as in the following statements:

```
%let wk=week;

title "This is data for &wk.1";
```

CAUTION:

Do not use AF, DMS, or SYS as prefixes with macro variable names. The letters AF, DMS, and SYS are frequently used by SAS as prefixes for macro variables created by SAS. SAS does not prevent you from using AF, DMS, or SYS as a prefix for macro variable names. However, using these strings as prefixes might create a conflict between the names that you specify and the name of a SAS created macro variable (including automatic macro variables in later SAS releases). If a name conflict occurs, SAS might not issue a warning or error message, depending on the details of the conflict. Therefore, the best practice is to avoid using the strings AF, DMS, or SYS as the beginning characters of macro names and macro variable names.

Solving Problems with Macro Variable Scope

A common mistake that occurs with macro variables concerns referencing local macro variables outside of their scopes. As described in "Scopes of Macro Variables" on page 43 macro variables are either global or local. Referencing a variable outside of its scope prevents the macro processor from resolving the variable reference. For example, consider the following program:

```
%macro totinv(var);
   data inv;
      retain total 0;
      set sasuser.houses end=final;
      total=total+&var;
      if final then call symput("macvar",put(total,dollar14.2));
   run;
```

```
   %put **** TOTAL=&macvar ****;
%mend totinv;

%totinv(price)
%put **** TOTAL=&macvar ****;      /* ERROR */
```

When you submit these statements, the %PUT statement in the macro TOTINV writes the value of TOTAL to the log. The %PUT statement that follows the macro call generates a warning message and writes the text **`TOTAL=&macvar`** to the log, as follows:

```
TOTAL= $1,240,800.00
WARNING: Apparent symbolic reference MACVAR not resolved.
**** TOTAL=&macvar ****
```

The second %PUT statement fails because the macro variable MACVAR is local to the TOTINV macro. To correct the error, you must use a %GLOBAL statement to declare the macro variable MACVAR.

Another common mistake that occurs with macro variables concerns overlapping macro variable names. If, within a macro definition, you refer to a macro variable with the same name as a global macro variable, you affect the global variable, which might not be what you intended. Either give your macro variables distinct names or use a %LOCAL statement to specifically define the variables in a local scope. See "Forcing a Macro Variable to Be Local" on page 58 for an example of this technique.

Solving Open Code Statement Recursion Problems

Recursion is something calling itself. *Open code recursion* is when your open code erroneously causes a macro statement to call another macro statement. This call is referred to as a recursive reference. The most common error that causes open code recursion is a missing semicolon. In the following example, the %LET statement is not terminated by a semicolon:

```
%let a=b   /* ERROR */
%put **** &a ****;
```

When the macro processor encounters the %PUT statement within the %LET statement, it generates this error message:

```
ERROR: Open code statement recursion detected.
```

Open code recursion errors usually occur because the macro processor is not reading your macro statements as you intended. Careful proofreading can usually solve open code recursion errors, because this type of error is mostly the result of typos in your code, not errors in execution logic.

To recover from an open code recursion error, first try submitting a single semicolon. If that does not work, try submitting the following string:

```
*'; *"; *); */; %mend; run;
```

Continue submitting this string until the following message appears in the SAS log:

```
ERROR: No matching %MACRO statement for this %MEND statement.
```

If the above method does not work, close your SAS session and restart SAS. Of course, closing and restarting SAS causes you to lose any unsaved data. Be sure to save often while you are developing your macros, and proofread them carefully before you submit them.

Solving Problems with Macro Functions

Some common causes of problems with macro functions include the following:

- misspelling the function name
- omitting the opening or closing parenthesis
- omitting an argument or specifying an extra argument

If you encounter an error related to a macro function, you might also see other error messages. The messages are generated by the invalid tokens left on the input stack by the macro processor.

Consider the following example. The user wants to use the %SUBSTR function to assign a portion of the value of the macro variable LINCOLN to the macro variable SECONDWD. But a typo exists in the second %LET statement, where %SUBSTR is misspelled as %SUBSRT:

```
%macro test;
%let lincoln=Four score and seven;
%let secondwd=%subsrt(&lincoln,6,5);   /* ERROR */
%put *** &secondwd ***;
%mend test;

%test
```

When the erroneous program is submitted, the following appears in the SAS log:

```
WARNING: Apparent invocation of macro SUBSRT not resolved.
```

The error messages clearly point to the function name, which is misspelled.

Solving Unresolved Macro Problems

When a macro name is passed to the macro processor but the processor does not find a matching macro definition, it generates the following message:

```
WARNING: Apparent invocation of macro
NAME not resolved.
```

This error could be caused by the following:

- the misspelling of the name of a macro or a macro function
- an error in a macro definition that caused the macro to be compiled as a dummy macro

A *dummy macro* is a macro that the macro processor partially compiles but does not store.

Note: You receive this warning only if the MERROR system option is on.

Solving the "Black Hole" Macro Problem

When the macro processor begins compiling a macro definition, it reads and compiles tokens until it finds a matching %MEND statement. If you omit a %MEND statement or cause it to be unrecognized by omitting a semicolon in the preceding statement, the macro processor does not stop compiling tokens. Every line of code that you submit becomes part of the macro.

Resubmitting the macro definition and adding the %MEND statement does not correct the error. When you submit the corrected definition, the macro processor treats it as a nested definition in the original macro definition. The macro processor must find a matching %MEND statement to stop compilation.

Note: It is a good practice to use the %MEND statement with the macro name, so you can easily match %MACRO and %MEND statements.

If you recognize that SAS is not processing submitted statements and you are not sure how to recover, submit %MEND statements one at a time until the following message appears in the SAS log:

```
ERROR: No matching %MACRO statement for this %MEND statement.
```

Then recall the original erroneous macro definition, correct the error in the %MEND statement, and submit the definition for compilation.

There are other syntax errors that can create similar problems, such as unmatched quotation marks and unclosed parentheses. Often, one of these syntax errors leads to others. Consider the following example:

```
%macro rooms;
   /* other macro statements& */
   %put **** %str(John's office) ****;    /* ERROR */
%mend rooms;

%rooms
```

When you submit these statements, the macro processor begins to compile the macro definition ROOMS. However, the single quotation mark in the %PUT statement is not marked by a percent sign. Therefore, during compilation the macro processor interprets the single quotation mark as the beginning of a literal token. It does not recognize the closing parenthesis, the semicolon at the end of the statement, or the %MEND statement at the end of the macro definition.

To recover from this error, you must submit the following:

```
');
%mend;
```

If the above methods do not work, try submitting the following string:

```
*'; *"; *); */; %mend; run;
```

Continue submitting this string until the following message appears in the SAS log:

```
ERROR: No matching %MACRO statement for this %MEND statement.
```

Obviously, it is easier to catch these errors before they occur. You can avoid subtle syntax errors by carefully checking your macros before submitting them for compilation. See "Developing Bug-free Macros" on page 121 for a syntax checklist.

Note: Another cause of unexplained and unexpected macro behavior is using a reserved word as the name of a macro variable or macro. For example, because SAS reserves names starting with SYS, you should not create macros and macro variables with names beginning with SYS. Most host environments have reserved words too. For example, on PC-based platforms, the word CON is reserved for console input. Check

" Reserved Words in the Macro Facility" on page 363 for reserved SAS keywords. Check your SAS companion for host environment reserved words.

Resolving Timing Issues

Many macro errors occur because a macro variable resolves at a different time than when the user intended or a macro statement executes at an unexpected time. A prime example of the importance of timing is when you use CALL SYMPUT to write a DATA step variable to a macro variable. You cannot use this macro variable in the same DATA step where it is defined; only in subsequent steps (after the DATA step's RUN statement).

The key to preventing timing errors is to understand how the macro processor works. In simplest terms, the two major steps are compilation and execution. The compilation step resolves all macro code to compiled code. Then the code is executed. Most timing errors occur because of the following:

- the user expects something to happen during compilation that does not actually occur until execution
- expects something to happen later but is actually executed right away

Here are two examples to help you understand why the timing of compilation and execution can be important.

Example of a Macro Statement Executing Immediately

In the following program, the user intends to use the %LET statement and the SR_CIT variable to indicate whether a data set contains any data for senior citizens:

```
data senior;
   set census;
   if age > 65 then
   do;
      %let sr_cit = yes;   /* ERROR */
      output;
   end;
run;
```

However, the results differ from the user's expectations. The %LET statement is executed immediately, while the DATA step is being compiled--before the data set is read. Therefore, the %LET statement executes regardless of the results of the IF condition. Even if the data set contains no observations where AGE is greater than 65, SR_CIT is always **yes**.

The solution is to set the macro variable's value by a means that is controlled by the IF logic and does not execute unless the IF statement is true. In this case, the user should use CALL SYMPUT, as in the following correct program:

```
%let sr_cit = no;
data senior;
   set census;
   if age > 65 then
   do;
      call symput ("sr_cit","yes");
    output;
   end;
run;
```

When this program is submitted, the value of SR_CIT is set to **yes**only if an observation is found with AGE greater than 65. Note that the variable was initialized to **no**. It is generally a good idea to initialize your macro variables.

Resolving Macro Resolution Problems Occurring During DATA Step Compilation

In the previous example, you learned you had to use CALL SYMPUT to conditionally assign a macro variable a value in a DATA step. So, you submit the following program:

```
%let sr_age = 0;
data senior;
   set census;
   if age > 65 then
   do;
      call symput("sr_age",age);
      put "This data set contains data about a person";
      put "who is &sr_age years old."; /* ERROR */
   end;
run;
```

If AGE was 67, you would expect to see a log message like the following:

```
This data set contains data about a person
who is 67 years old.
```

However, no matter what AGE is, the following message is sent to the log:

```
This data set contains data about a person
who is 0 years old.
```

When the DATA step is being compiled, &SR_AGE is sent to the macro facility for resolution, and the result is passed back before the DATA step executes. To achieve the desired result, submit this corrected program instead:

```
%let sr_age = 0;
data senior;
   set census;
   if age > 65 then
   do;
      call symput("sr_age",age);
      stop;
   end;
run;

data _null_;
   put "This data set contains data about a person";
   put "who is &sr_age years old.";
run;
```

Note: Use double quotation marks in statements like PUT, because macro variables do not resolve when enclosed in single quotation marks.

Here is another example of erroneously referring to a macro variable in the same step that creates it:

```
data _null_;
   retain total 0;
   set mydata end=final;
   total=total+price;
```

```
    call symput("macvar",put(total,dollar14.2));
    if final then put "*** total=&macvar ***"; /* ERROR */
run;
```

When these statements are submitted, the following lines are written to the SAS log:

```
WARNING: Apparent symbolic reference MACVAR not resolved.

*** total=&macvar ***
```

As this DATA step is tokenized and compiled, the **&**causes the word scanner to trigger the macro processor, which looks for a MACVAR entry in a symbol table. Because such an entry does not exist, the macro processor generates the warning message. Because the tokens remain on the input stack, they are transferred to the DATA step compiler. During DATA step execution, the CALL SYMPUT statement creates the macro variable MACVAR and assigns a value to it. However, the text **&macvar**in the PUT statement occurs because the text has already been processed while the macro was being compiled. If you were to resubmit these statements and the macro would appear to work correctly, but the value of MACVAR would reflect the value set during the previous execution of the DATA step. This value can be misleading.

Remember that in general, the **%** and **&**trigger immediate execution or resolution during the compilation stage of the rest of your SAS code.

For more examples and explanation of how CALL SYMPUT creates macro variables, see "Special Cases of Scope with the CALL SYMPUT Routine" on page 63.

Solving Problems with the Autocall Facility

The autocall facility is an efficient way of storing and using production (debugged) macros. When a call to an autocall macro produces an error, the cause is one of two things:

- an erroneous autocall library specification
- an invalid autocall macro definition.

If the error is the autocall library specification and the MERROR option is set, SAS can generate any or all of the following warnings:

```
WARNING: No logical assign for filename
FILENAME.
WARNING: Source level autocall is not found or cannot be opened.
         Autocall has been suspended and OPTION NOMAUTOSOURCE has
         been set. To use the autocall facility again, set OPTION
         MAUTOSOURCE.
WARNING: Apparent invocation of macro
MACRO-NAME not resolved.
```

If the error is in the autocall macro definition, SAS generates a message like the following:

```
NOTE: Line generated by the invoked macro
"MACRO-NAME".
```

Fixing Autocall Library Specifications

When an autocall library specification causes an error, it is because the macro processor cannot find the member containing the autocall macro definition in the library or libraries specified in the SASAUTOS system option.

To correct this error, follow these steps.

1. If the unresolved macro call created an invalid SAS statement, submit a single semicolon to terminate the invalid statement. SAS is then able to correctly recognize subsequent statements.
2. Look at the value of the SASAUTOS system option by printing the output of the OPTIONS procedure or by viewing the OPTIONS window in the SAS windowing environment. (Or, edit your SAS configuration file or SAS autoexec file.) Verify each fileref or directory name. If you find an error, submit a new OPTIONS statement or change the SASAUTOS setting in the OPTIONS window.
3. Check the MAUTOSOURCE system option. If SAS could not open at least one library, it sets the NOMAUTOSOURCE option. If NOMAUTOSOURCE is present, reset MAUTOSOURCE with a new OPTIONS statement or the OPTIONS window.
4. If the library specifications are correct, check the contents of each directory to verify that the autocall library member exists and that it contains a macro definition of the same name. If the member is missing, add it.
5. Set the MRECALL option with a new OPTIONS statement or the OPTIONS window. By default, the macro processor searches only once for an undefined macro. Setting this option causes the macro processor to search the autocall libraries for the specification again.
6. Call the autocall macro, which includes and submits the autocall macro source.
7. Reset the NOMRECALL option.

Note: Some host environments have environment variables or system-level logical names assigned to the SASAUTOS library; check your SAS companion for more information about how the SASAUTOS library specification is handled in your host environment.

Fixing Autocall Macro Definition Errors

When the autocall facility locates an autocall library member, the macro processor compiles any macros in that library member. It stores the compiled macros in the catalog containing stored compiled macros. For the rest of your SAS session, invoking one of those macros retrieves the compiled macro from the WORK library. Under no circumstances does the autocall facility use an autocall library member when a compiled macro with the same name already exists. Thus, if you invoke an autocall macro and discover you made an error when you defined it, you must correct the autocall library member for future use. Compile the corrected version directly in your program or session.

To correct an autocall macro definition in a windowing environment, do the following:

1. Use the INCLUDE command to bring the autocall library member into the SAS Program Editor window. If the macro is stored in a catalog SOURCE entry, use the COPY command to bring the program into the Program Editor window.

2. Correct the error.
3. Store a copy of the corrected macro in the autocall library with the FILE command for a macro in an external file or with a SAVE command for a macro in a catalog entry.
4. Submit the macro definition from the Program Editor window.

The macro processor then compiles the corrected version, replacing the incorrect compiled macro. The corrected, compiled macro is now ready to execute at the next invocation.

To correct an autocall macro definition in an interactive line mode session, do the following:

1. Edit the autocall macro source with a text editor.
2. Correct the error.
3. Use a %INCLUDE statement to bring the corrected library member into your SAS session.

The macro processor then compiles the corrected version, replacing the incorrect compiled macro. The corrected, compiled macro is now ready to execute at the next invocation.

File and Macro Names for Autocall

When you want to use a macro as an autocall macro, you must store the macro in a file with the same name as the macro. Also, the file extension must be **.sas** (if your operating system uses file extensions). If you experience problems with the autocall facility, be sure the macro and filenames match and the file has the right extension when necessary.

Displaying Information about Stored Compiled Macros

To display the list of entries in a catalog containing compiled macros, you can use the Catalog window or the CATALOG procedure. The following PROC step displays the contents of a macro catalog in a SAS library identified with the libref MYSASLIB:

```
libname mysaslib
'SAS-data-library';
   proc catalog catalog=mysaslib.sasmacr;
      contents;
   run;
   quit;
```

You can also use PROC CATALOG to display information about autocall library macros stored in SOURCE entries in a catalog. You cannot use PROC CATALOG or the Explorer window to copy, delete, or rename stored compiled macros.

You can use the MCOMPILENOTE system option to issue a note to the log upon the completion of the compilation of any macro. For more information, see “MCOMPILENOTE System Option” on page 335.

In SAS 6.11 and later, you can use PROC SQL to retrieve information about all compiled macros. For example, submitting these statements produces output similar to the following output:

```
proc sql;
   select * from dictionary.catalogs
        where memname in ('SASMACR');
```

Output 10.1 *Output from PROC SQL Program for Viewing Compiled Macros*

```
Library  Member   Member   Object   Object                         Date      Object
Name     Name     Type     Name     Type    Object Description     Modified  Alias
------------------------------------------------------------------------------------
WORK     SASMACR  CATALOG  FINDAUTO MACRO                          05/28/96

SASDATA  SASMACR  CATALOG  CLAUSE   MACRO   Count words in clause  05/24/96

SASDATA  SASMACR  CATALOG  CMPRES   MACRO   CMPRES autocall macro  05/24/96

SASDATA  SASMACR  CATALOG  DATATYP  MACRO   DATATYP autocall macro 05/24/96

SASDATA  SASMACR  CATALOG  LEFT     MACRO   LEFT autocall macro    05/24/96
```

To display information about compiled macros when you invoke them, use the SAS system options MLOGIC, MPRINT, and SYMBOLGEN. When you specify the SAS system option MLOGIC, the libref and date of compilation of a stored compiled macro are written to the log along with the usual information displayed during macro execution.

Solving Problems with Expression Evaluation

The following macro statements use the %EVAL function:

Table 10.3 *Macro Statements That Use the %EVAL Function*

%DO	%IF-%THEN	%SCAN
%DO %UNTIL	%QSCAN	%SYSEVALF
%DO %WHILE	%QSUBSTR	%SUBSTR

In addition, you can use the %EVAL function to specify an expression evaluation.

The most common errors that occur while evaluating expressions are the presence of character operands where numeric operands are required or ambiguity about whether a token is a numeric operator or a character value. "Macro Expressions" on page 71 discusses these and other macro expression errors.

Quite often, an error occurs when a special character or a keyword appears in a character string. Consider the following program:

```
%macro conjunct(word= );
   %if &word = and or &word = but or &word = or %then   /* ERROR */
      %do %put *** &word is a conjunction. ***;

   %else
      %do %put *** &word is not a conjunction. ***;
%mend conjunct;
```

In the %IF statement, the values of WORD being tested are ambiguous — they could also be interpreted as the numeric operators AND and OR. Therefore, SAS generates the following error messages in the log:

```
ERROR: A character operand was found in the %EVAL function or %IF
       condition where a numeric operand is required. The condition
       was:word = and or       &word = but or       &word = or
ERROR: The macro will stop executing.
```

To fix this problem, use the quoting functions %BQUOTE and %STR, as in the following corrected program:

```
%macro conjunct(word= );
   %if %bquote(&word) = %str(and) or %bquote(&word) = but or
         %bquote(&word) = %str(or) %then
      %do %put *** &word is a conjunction. ***;

   %else
      %do %put *** &word is not a conjunction. ***;
%mend conjunct;
```

In the corrected program, the %BQUOTE function quotes the result of the macro variable resolution (in case the user passes in a word containing an unmatched quotation mark or some other odd value). The %STR function quotes the comparison values AND and OR at compile time, so they are not ambiguous. You do not need to use %STR on the value BUT, because it is not ambiguous (not part of the SAS or macro languages). See “Macro Quoting” on page 80 for more information about using macro quoting functions.

Debugging Techniques

Using System Options to Track Problems

The SAS system options MLOGIC, MLOGICNEST, MPRINT, MPRINTNEST, and SYMBOLGEN can help you track the macro code and SAS code generated by your macro. Messages generated by these options appear in the SAS log, prefixed by the name of the option responsible for the message.

Note: Whenever you use the macro facility, use the following macro options: MACRO, MERROR, and SERROR. SOURCE is a system option that is helpful when using the macro facility. It is also helpful to use the SOURCE2 system option when using the %INCLUDE.

Although the following sections discuss each system option separately, you can, of course, combine them. However, each option can produce a significant amount of output, and too much information can be as confusing as too little. So, use only those options that you think you might need and turn them off when you complete the debugging.

Tracing the Flow of Execution with MLOGIC

The MLOGIC system option traces the flow of execution of your macro, including the resolution of parameters, the scope of variables (global or local), the conditions of macro expressions being evaluated, the number of loop iterations, and the beginning and end of each macro execution. Use the MLOGIC option when you think a bug lies in the program logic (as opposed to simple syntax errors).

Note: MLOGIC can produce a lot of output, so use it only when necessary, and turn it off when debugging is finished.

In the following example, the macro FIRST calls the macro SECOND to evaluate an expression:

```
%macro second(param);
   %let a = %eval(&param); &a
%mend second;

%macro first(exp);
   %if (%second(&exp) ge 0) %then
      %put **** result >= 0 ****;
   %else
      %put **** result < 0 ****;
%mend first;

options mlogic;
%first(1+2)
```

Submitting this example with option MLOGIC shows when each macro starts execution, the values of passed parameters, and the result of the expression evaluation.

```
MLOGIC(FIRST):  Beginning execution.
MLOGIC(FIRST):  Parameter EXP has value 1+2
MLOGIC(SECOND):  Beginning execution.
MLOGIC(SECOND):  Parameter PARAM has value 1+2
MLOGIC(SECOND):  %LET (variable name is A)
MLOGIC(SECOND):  Ending execution.
MLOGIC(FIRST):  %IF condition (%second(&exp) ge 0) is TRUE
MLOGIC(FIRST):  %PUT **** result >= 0 ****
MLOGIC(FIRST):  Ending execution.
```

Nesting Information Generated by MLOGICNEST

MLOGICNEST allows the macro nesting information to be written to the SAS log in the MLOGIC output. The setting of MLOGICNEST does not imply the setting of MLOGIC. You must set both MLOGIC and MLOGICNEST in order for output (with nesting information) to be written to the SAS log.

For more information and an example, see “MLOGICNEST System Option” on page 347.

Examining the Generated SAS Statements with MPRINT

The MPRINT system option writes to the SAS log each SAS statement generated by a macro. Use the MPRINT option when you suspect your bug lies in code that is generated in a manner that you did not expect.

For example, the following program generates a simple DATA step:

```
%macro second(param);
   %let a = %eval(&param); &a
%mend second;
```

```
%macro first(exp);
   data _null_;
      var=%second(&exp);
      put var=;
   run;
%mend first;

options mprint;
%first(1+2)
```

When you submit these statements with option MPRINT, these lines are written to the SAS log:

```
MPRINT(FIRST):   DATA _NULL_;
MPRINT(FIRST):   VAR=
MPRINT(SECOND):  3
MPRINT(FIRST):  ;
MPRINT(FIRST):   PUT VAR=;
MPRINT(FIRST):   RUN;

VAR=3
```

The MPRINT option shows you the generated text and identifies the macro that generated it.

Nesting Information Generated by MPRINTNEST

MPRINTNEST allows the macro nesting information to be written to the SAS log in the MPRINT output. This value has no effect on the MPRINT output that is sent to an external file. For more information, see "MFILE System Option" on page 343.

The setting of MPRINTNEST does not imply the setting of MPRINT. You must set both MPRINT and MPRINTNEST in order for output (with the nesting information) to be written to the SAS log.

For more information and an example, see "MPRINTNEST System Option" on page 351.

Storing MPRINT Output in an External File

You can store text that is generated by the macro facility during macro execution in an external file. Printing the statements generated during macro execution to a file is useful for debugging macros when you want to test generated text in a later SAS session.

To use this feature, set both the MFILE and MPRINT system options on. Also assign MPRINT as the fileref for the file to contain the output generated by the macro facility:

```
options mprint mfile;
filename mprint 'external-file';
```

The external file created by the MPRINT system option remains open until the SAS session terminates. The MPRINT text generated by the macro facility is written to the log during the SAS session and to the external file when the session ends. The text consists of program statements generated during macro execution with macro variable references and macro expressions resolved. Only statements generated by the macro are stored in the external file. Any program statements outside the macro are not written to the external file. Each statement begins on a new line with one space separating words.

The text is stored in the external file without the **MPRINT(*macroname*:** prefix, which is displayed in the log.

If MPRINT is not assigned as a fileref or if the file cannot be accessed, warnings are written to the log and MFILE is turned off. To use the feature again, you must specify MFILE again.

By default, the MPRINT and MFILE options are off.

The following example uses the MPRINT and MFILE options to store generated text in the external file named TEMPOUT:

```
options mprint mfile;
filename mprint 'TEMPOUT';

%macro temp;
   data one;
      %do i=1 %to 3;
         x&i=&i;
      %end;
   run;
%mend temp;

%temp
```

The macro facility writes the following lines to the SAS log and creates the external file named TEMPOUT:

```
MPRINT(TEMP):   DATA ONE;
NOTE: The macro generated output from MPRINT will also be written
      to external file '/u/local/abcdef/TEMPOUT' while OPTIONS
      MPRINT and MFILE are set.
MPRINT(TEMP):   X1=1;
MPRINT(TEMP):   X2=2;
MPRINT(TEMP):   X3=3;
MPRINT(TEMP):   RUN;
```

When the SAS session ends, the file TEMPOUT contains:

```
DATA ONE;
X1=1;
X2=2;
X3=3;
RUN;
```

Note: Using MPRINT to write code to an external file is a debugging tool only. It should not be used to create SAS code files for purposes other than debugging.

Examining Macro Variable Resolution with SYMBOLGEN

The SYMBOLGEN system option tells you what each macro variable resolves to by writing messages to the SAS log. This option is especially useful in spotting quoting problems, where the macro variable resolves to something other than what you intended because of a special character.

For example, suppose you submit the following statements:

```
options symbolgen;

%let a1=dog;
```

```
%let b2=cat;
%let b=1;
%let c=2;
%let d=a;
%let e=b;
%put **** &&&d&b ****;
%put **** &&&e&c ****;
```

The SYMBOLGEN option writes these lines to the SAS log:

```
SYMBOLGEN:  && resolves to &.
SYMBOLGEN:  Macro variable D resolves to a
SYMBOLGEN:  Macro variable B resolves to 1
SYMBOLGEN:  Macro variable A1 resolves to dog
**** dog ****

SYMBOLGEN:  && resolves to &.
SYMBOLGEN:  Macro variable E resolves to b
SYMBOLGEN:  Macro variable C resolves to 2
SYMBOLGEN:  Macro variable B2 resolves to cat
**** cat ****
```

Reading the log provided by the SYMBOLGEN option is easier than examining the program statements to trace the indirect resolution. Notice that the SYMBOLGEN option traces each step of the macro variable resolution by the macro processor. When the resolution is complete, the %PUT statement writes the value to the SAS log.

When you use SYMBOLGEN to trace the values of macro variables that have been masked with a macro quoting function, you might see an additional message about the quoting being "stripped for printing." For example, suppose you submit the following statements, with SYMBOLGEN set to on:

```
%let nickname = %str(My name%'s O%'Malley, but I%'m called Bruce);
%put *** &nickname ***;
```

The SAS log contains the following after these statements have executed:

```
SYMBOLGEN:  Macro variable NICKNAME resolves to
                            My name's O'Malley, but I'm called Bruce
SYMBOLGEN:  Some characters in the above value which were
                           subject to macro quoting have been
                           unquoted for printing.
*** My name's O'Malley, but I'm called Bruce ***
```

You can ignore the unquoting message.

Using the %PUT Statement to Track Problems

Along with using the SYMBOLGEN system option to write the values of macro variables to the SAS log, you might find it useful to use the %PUT statement while developing and debugging your macros. When the macro is finished, you can delete or comment out the %PUT statements. The following table provides some occasions where you might find the %PUT statement helpful in debugging, and an example of each:

Table 10.4 *Example %PUT Statements That Are Useful when Debugging Macros*

Situation	Example
show a macro variable's value	`%PUT ****&=variable-name****;`
check leading or trailing blanks in a variable's value	`%PUT ***&variable-name***;`
check double-ampersand resolution, as during a loop	`%PUT ***variable-name&i = &&variable-name***;`
check evaluation of a condition	`%PUT ***This condition was met.***;`

As you recall, macro variables are stored in symbol tables. There is a global symbol table, which contains global macro variables, and a local symbol table, which contains local macro variables. During the debugging process, you might find it helpful on occasion to print these tables to examine the scope and values of a group of macro variables. To do so, use the %PUT statement with one of the following options:

_ALL
: describes all currently defined macro variables, regardless of scope. User-generated global and local variables as well as automatic macro variables are included.

AUTOMATIC
: describes all automatic macro variables. The scope is listed as AUTOMATIC. All automatic macro variables are global except SYSPBUFF.

GLOBAL
: describes all global macro variables that were not created by the macro processor. The scope is listed as GLOBAL. Automatic macro variables are not listed.

LOCAL
: describes user-generated local macro variables defined within the currently executing macro. The scope is listed as the name of the macro in which the macro variable is defined.

USER
: describes all user-generated macro variables, regardless of scope. For global macro variables, the scope is GLOBAL; for local macro variables, the scope is the name of the macro.

The following example uses the %PUT statement with the argument _USER_ to examine the global and local variables available to the macro TOTINV. Notice the use of the user-generated macro variable TRACE to control when the %PUT statement writes values to the log.

```
%macro totinv(var);
   %global macvar;
   data inv;
      retain total 0;
      set sasuser.houses end=final;
      total=total+&var;
      if final then call symput("macvar",put(total,dollar14.2));
   run;
```

```
    %if &trace = ON  %then
      %do;
         %put *** Tracing macro scopes. ***;
         %put _USER_;
      %end;
%mend totinv;

%let trace=ON;
%totinv(price)
%put *** TOTAL=&macvar ***;
```

When you submit these statements, the first %PUT statement in the macro TOTINV writes the message about tracing being on and then writes the scope and value of all user generated macro variables to the SAS log.

```
*** Tracing macro scopes. ***
TOTINV VAR price
GLOBAL TRACE ON
GLOBAL MACVAR  $1,240,800.00
*** TOTAL= $1,240,800.00 ***
```

See “Scopes of Macro Variables” on page 43 for a more detailed discussion of macro variable scopes.

Chapter 11
Writing Efficient and Portable Macros

Writing Efficient and Portable Macros

The macro facility is a powerful tool for making your SAS code development more efficient. But macros are only as efficient as you make them. There are several techniques and considerations for writing efficient macros. If you intend to extend the power of the macro facility by creating macros that can be used on more than one host environment, there are additional considerations for writing portable macros.

Keeping Efficiency in Perspective

Efficiency is an elusive thing, hard to quantify and harder still to define. What works with one application might not work with another, and what is efficient on one host environment might be inefficient on a different system. However, there are some generalities that you should keep in mind.

Usually, efficiency issues are discussed in terms of CPU cycles, elapsed time, I/O hits, memory usage, disk storage, and so on. This section does not give benchmarks in these terms because of all the variables involved. A program that runs only once needs different tuning than a program that runs hundreds of times. An application running on a mainframe has different hardware parameters than an application developed on a desktop PC. You must keep efficiency in perspective with your environment.

There are different approaches to efficiency, depending on what resources you want to conserve. Are CPU cycles more critical than I/O hits? Do you have lots of memory but no disk space? Taking stock of your situation before deciding how to tune your programs is a good idea.

The area of efficiency most affected by the SAS macro facility is human efficiency — how much time is required to both develop and maintain a program. Autocall macros are particularly important in this area because the autocall facility provides code reusability. Once you develop a macro that performs a task, you can save it and use it not only in the application that you developed it for, but also in future applications without any further work. A library of reusable, immediately callable macros is a boon to any application development team.

The stored compiled macro facility (described in "Storing and Reusing Macros" on page 113) might reduce execution time by enabling previously compiled macros to be accessed during different SAS jobs and sessions. But it is a tool that is efficient only for production applications, not during application development. So the efficiency techniques that you choose depend not only on your hardware and personnel situation, but also on the stage that you have reached in your application development process.

Also, remember that incorporating macro code into a SAS application does not automatically make the application more efficient. When designing a SAS application, concentrate on making the basic SAS code that macros generate more efficient. There are many sources for information about efficient SAS code, including *SAS Programming Tips: A Guide to Efficient SAS Processing.*

Writing Efficient Macros

Use Macros Wisely

An application that uses a macro to generate only constant text is inefficient. In general, for these situations consider using a %INCLUDE statement. Because the %INCLUDE statement does not have to compile the code first (it is executed immediately), it might be more efficient than using a macro (especially if the code is executed only once). If you use the same code repeatedly, it might be more efficient to use a macro because a macro is compiled only once during a SAS job, no matter how many times it is called.

However, using %INCLUDE requires you to know exactly where the physical file is stored and specify this name in the program itself. Because with the autocall facility all you have to remember is the name of the macro (not a pathname), the gain in human efficiency might more than offset the time gained by not compiling the macro. Also, macros provide additional programming features, such as parameters, conditional sections, and loops, as well as the ability to view macro variable resolution in the SAS log.

So, be sure to use a macro only when necessary. And, balance the various efficiency factors and gains (how many times you use the code, CPU time versus ease-of-use) to reach a solution that is best for your application.

Use Name Style Macros

Macros come in three invocation types: name style, command style, and statement style. Of the three, name style is the most efficient because name style macros always begin with a `%`, which immediately tells the word scanner to pass the token to the macro processor. With the other two types, the word scanner does not know immediately whether the token should be sent to the macro processor. Therefore, time is wasted while the word scanner determines whether the token should be sent.

Avoid Nested Macro Definitions

Nesting macro definitions inside other macros is usually unnecessary and inefficient. When you call a macro that contains a nested macro definition, the macro processor generates the nested macro definition as text and places it on the input stack. The word scanner then scans the definition and the macro processor compiles it. If you nest the definition of a macro that does not change, you cause the macro processor to compile the same macro each time that section of the outer macro is executed.

As a rule, you should define macros separately. If you want to nest a macro's scope, simply nest the macro call, not the macro definition.

For example, the macro STATS1 contains a nested macro definition for the macro TITLE:

```
/* Nesting a Macro Definition--INEFFICIENT */
%macro stats1(product,year);
   %macro title;
      title "Statistics for &product in &year";
      %if &year>1929 and &year<1935 %then
         %do;
            title2 "Some Data Might Be Missing";
         %end;
   %mend title;

   proc means data=products;
      where product="&product" and year=&year;
      %title
   run;
%mend stats1;

%stats1(steel,2002)
%stats1(beef,2000)
%stats1(fiberglass,2001)
```

Each time the macro STATS1 is called, the macro processor generates the definition of the macro TITLE as text, recognizes a macro definition, and compiles the macro TITLE. In this case, STATS1 was called three times, which means the TITLE macro was compiled three times. With only a few statements, this task takes only micro-seconds; but in large macros with hundreds of statements, the wasted time could be significant.

The values of PRODUCT and YEAR are available to TITLE because its call is within the definition of STATS1. Therefore, it is unnecessary to nest the definition of TITLE to make values available to TITLE's scope. Nesting definitions are also unnecessary because no values in the definition of the TITLE statement are dependent on values that change during the execution of STATS1. (Even if the definition of the TITLE statement depended on such values, you could use a global macro variable to effect the changes, rather than nest the definition.)

The following program shows the macros defined separately:

```
/* Separating Macro Definitions--EFFICIENT */
%macro stats2(product,year);
   proc means data=products;
      where product="&product" and year=&year;
      %title
   run;
%mend stats2;

%macro title;
   title "Statistics for &product in &year";
   %if &year>1929 and &year<1935 %then
      %do;
         title2 "Some Data Might Be Missing";
      %end;
%mend title;

%stats2(cotton,1999)
%stats2(brick,2002)
%stats2(lamb,2001)
```

Here, because the definition of the macro TITLE is outside the definition of the macro STATS2, TITLE is compiled only once, even though STATS2 is called three times. Again, the values of PRODUCT and YEAR are available to TITLE because its call is within the definition of STATS2.

Note: Another reason to define macros separately is because it makes them easier to maintain, each in a separate file.

Assign Function Results to Macro Variables

It is more efficient to resolve a variable reference than it is to evaluate a function. Therefore, assign the results of frequently used functions to macro variables.

For example, the following macro is inefficient because the length of the macro variable THETEXT must be evaluated at every iteration of the %DO %WHILE statement:

```
/* INEFFICIENT MACRO */
%macro test(thetext);
   %let x=1;
      %do %while (&x > %length(&thetext));
         .
         .
```

```
         .
      %end;
%mend test;

%test(Four Score and Seven Years Ago)
```

A more efficient method would be to evaluate the length of THETEXT once and assign that value to another macro variable. Then, use that variable in the %DO %WHILE statement, as in the following program:

```
/* MORE EFFICIENT MACRO */
%macro test2(thetext);
   %let x=1;
   %let length=%length(&thetext);
   %do %while (&x > &length);
      .
      .
      .
   %end;
%mend test2;

%test(Four Score and Seven Years Ago)
```

As another example, suppose you want to use the %SUBSTR function to pull the year out of the value of SYSDATE. Instead of using %SUBSTR repeatedly in your code, assign the value of the %SUBSTR(&SYSDATE, 6) to a macro variable, then use that variable whenever you need the year.

Turn Off System Options When Appropriate

While the debugging system options, such as MPRINT and MLOGIC, are very helpful at times, it is inefficient to run production (debugged) macros with this type of system option set to on. For production macros, run your job with the following settings: NOMLOGIC, NOMPRINT, NOMRECALL, and NOSYMBOLGEN.

Even if your job has no errors, if you run it with these options turned on you incur the overhead that the options require. By turning them off, your program runs more efficiently.

Note: Another approach to deciding when to use MPRINT versus NOMPRINT is to match this option's setting with the setting of the SOURCE option. That is, if your program uses the SOURCE option, it should also use MPRINT. If your program uses NOSOURCE, then run it with NOMPRINT as well.

Note: If you do not use autocall macros, use the NOMAUTOSOURCE system option. If you do not use stored compiled macros, use the NOMSTORED system option.

Use the Stored Compiled Macro Facility

The stored compiled macro facility reduces execution time by enabling macros compiled in a previous SAS job or session to be accessed during subsequent SAS jobs and sessions. Therefore, these macros do not need to be recompiled. Use the stored compiled macro facility only for production (debugged) macros. It is not efficient to use this facility when developing a macro application.

CAUTION:

Save the source code. Because you cannot re-create the source code for a macro from the compiled code, you should keep a copy of the source code in a safe place, in case the compiled code becomes corrupted for some reason. Having a copy of the source is also necessary if you intend to modify the macro at a later time.

See “Storing and Reusing Macros” on page 113 for more information about the stored compiled macro facility.

Note: The compiled code generated by the stored compiled macro facility is not portable. If you need to transfer macros to another host environment, you must move the source code and recompile and store it on the new host.

Centrally Store Autocall Macros

When using the autocall facility, it is most efficient in terms of I/O to store all your autocall macros in one library and append that library name to the beginning of the SASAUTOS system option specification. Of course, you could store the autocall macros in as many libraries as you want--but each time you call a macro, each library is searched sequentially until the macro is found. Opening and searching only one library reduces the time SAS spends looking for macros.

However, it might make more sense, if you have hundreds of autocall macros, to have them separated into logical divisions according to purpose, levels of production, who supports them, and so on. As usual, you must balance reduced I/O against ease-of-use and ease-of-maintenance.

All autocall libraries in the concatenated list are opened and left open during a SAS job or session. The first time you call an autocall macro, any library that did not open the first time is tested again each time an autocall macro is used. Therefore, it is extremely inefficient to have invalid pathnames in your SASAUTOS system option specification. You see no warnings about this wasted effort on the part of SAS, unless no libraries at all open.

There are two efficiency tips involving the autocall facility:

- Do not store nonmacro code in autocall library files.
- Do not store more than one macro in each autocall library file.

Although these two practices are used by SAS and do work, they contribute significantly to code-maintenance effort and therefore are less efficient.

Other Useful Efficiency Tips

Here are some other efficiency techniques that you can try:

- Reset macro variables to null if the variables are no longer going to be referenced.
- Use triple ampersands to force an additional scan of macro variables with long values, when appropriate. See “Storing Only One Copy of a Long Macro Variable Value” on page 147 for more information.
- Adjust the values of the “MSYMTABMAX= System Option” on page 354 and “MVARSIZE= System Option” on page 355 to fit your situation. In general, increase the values if disk space is in short supply; decrease the values if memory is in short supply. MSYMTABMAX affects the space available for storing macro variable symbol tables; MVARSIZE affects the space available for storing values of individual macro variables.

Storing Only One Copy of a Long Macro Variable Value

Because macro variables can have very long values, the way you store macro variables can affect the efficiency of a program. Indirect references using three ampersands enable you to store fewer copies of a long value.

For example, suppose your program contains long macro variable values that represent sections of SAS programs:

```
%let pgm=%str(data flights;
   set schedule;
   totmiles=sum(of miles1-miles20);
   proc print;
   var flightid totmiles;);
```

You want the SAS program to end with a RUN statement:

```
%macro check(val);
      /* first version */    &val
    %if %index(&val,%str(run;))=0 %then %str(run;);
%mend check;
```

First, the macro CHECK generates the program statements contained in the parameter VAL (a macro variable that is defined in the %MACRO statement and passed in from the macro call). Then, the %INDEX function searches the value of VAL for the characters **run;**. (The %STR function causes the semicolon to be treated as text.) If the characters are not present, the %INDEX function returns 0. The %IF condition becomes true, and the macro processor generates a RUN statement.

To use the macro CHECK with the variable PGM, assign the parameter VAL the value of PGM in the macro call:

```
%check(&pgm)
```

As a result, SAS sees the following statements:

```
data flights;
   set schedule;
   totmiles=sum(of miles1-miles20);

proc print;
   var flightid totmiles;
run;
```

The macro CHECK works properly. However, the macro processor assigns the value of PGM as the value of VAL during the execution of CHECK. Thus, the macro processor must store two long values (the value of PGM and the value of VAL) while CHECK is executing.

To make the program more efficient, write the macro so that it uses the value of PGM rather than copying the value into VAL:

```
%macro check2(val);  /* more efficient macro */   &&&val
   %if %index(&&&val,%str(run;))=0 %then %str(run;);
%mend check2;

%check2(pgm)
```

The macro CHECK2 produces the same result as the macro CHECK:

```
data flights;
   set schedule;
   totmiles=sum(of miles1-miles20);

proc print;
   var flightid totmiles;
run;
```

However, in the macro CHECK2, the value assigned to VAL is simply the name **PGM**, not the value of PGM. The macro processor resolves &&&VAL into &PGM and then into the SAS statements contained in the macro variable PGM. Thus, the long value is stored only once.

Writing Portable Macros

Using Portable SAS Language Functions with %SYSFUNC

If your code runs in two different environments, you have essentially doubled the worth of your development effort. But portable applications require some planning ahead. For more details about any host-specific feature of SAS, see the SAS documentation for your host environment.

You can use the %SYSFUNC macro function to access SAS language functions to perform most host-specific operations, such as opening or deleting a file. For more information, see "%SYSFUNC and %QSYSFUNC Functions" on page 268.

Using %SYSFUNC to access portable SAS language functions can save you a lot of macro coding (and is therefore not only portable but also more efficient). The following table lists some common host-specific tasks and the functions that perform those tasks.

Table 11.1 *Portable SAS Language Functions and Their Uses*

Task	SAS Language Function or Functions
Assign and verify existence of fileref and physical file	FILENAME, FILEREF, PATHNAME
Open a file	FOPEN, MOPEN
Verify existence of a file	FEXIST, FILEEXIST
Get information about a file	FINFO, FOPTNAME, FOPTNUM
Write data to a file	FAPPEND, FWRITE
Read from a file	FPOINT, FREAD, FREWIND, FRLEN
Close a file	FCLOSE

Task	SAS Language Function or Functions
Delete a file	FDELETE
Open a directory	DOPEN
Return information about a directory	DINFO, DNUM, DOPTNAME, DOPTNUM, DREAD
Close a directory	DCLOSE
Read a host-specific option	GETOPTION
Interact with the File Data Buffer (FDB)	FCOL, FGET, FNOTE, FPOS, FPUT, FSEP
Assign and verify librefs	LIBNAME, LIBREF, PATHNAME
Get information about executed host environment commands	SYSRC

Note: Of course, you can also use other functions, such as ABS, MAX, and TRANWRD, with %SYSFUNC. A few SAS language functions are not available with %SYSFUNC. See "%SYSFUNC and %QSYSFUNC Functions" on page 268 for more details.

Example Using %SYSFUNC

The following program deletes the file identified by the fileref MYFILE:

```
%macro testfile(filrf);
   %let
rc=%sysfunc(filename(filrf,physical-filename));
   %if &rc = 0 and %sysfunc(fexist(&filrf)) %then
      %let rc=%sysfunc(fdelete(&filrf));
   %let rc=%sysfunc(filename(filrf));
%mend testfile;

%testfile(myfile)
```

Using Automatic Variables with Host-Specific Values

Macro Variables by Task

The automatic macro variables are available under all host environments, but the values are determined by each host. The following table lists the macro variables by task. The "Type" column tells you if the variable can be changed (Read and Write) or can be inspected (Read Only).

Table 11.2 Automatic Macro Variables with Host-Specific Results

Task	Automatic Macro Variable	Type
List the name of the current graphics device on DEVICE=.	SYSDEVIC	Read and write
List of the mode of execution (values are FORE or BACK). Some host environments allow only one mode, FORE.	SYSENV	Read-only
List the name of the currently executing batch job, user ID, or process. For example, on UNIX, SYSJOBID is the PID.	SYSJOBID	Read-only
List the last return code generated by your host environment, based on commands executed using the X statement in open code, the X command in the SAS windowing environment, or the %SYSEXEC (or %TSO or %CMS) macro statements. The default value is 0.	SYSRC	Read and write
List the abbreviation of the host environment that you are using.	SYSSCP	Read-only
List a more detailed abbreviation of the host environment that you are using.	SYSSCPL	Read-only
Retrieve a character string that was passed to SAS by the SYSPARM= system option.	SYSPARM	Read and write

Examples Using SYSSCP and SYSSCPL

The macro DELFILE uses the value of SYSSCP to determine the platform that is running SAS and deletes a TMP file. FILEREF is a macro parameter that contains a filename. Because the filename is host-specific, making it a macro parameter enables the macro to use whatever filename syntax is necessary for the host environment.

```
%macro delfile(fileref);
    /* Unix */
   %if &sysscp=HP 800 or &sysscp=HP 300 %then %do;
        X "rm &fileref..TMP";
   %end;

      /* DOS-LIKE platforms */
   %else %if &sysscp=OS2 or &sysscp=WIN %then %do;
            X "DEL &fileref..TMP";
   %end;
    /* CMS */
   %else %if &sysscp=CMS %then %do;
           X "ERASE &fileref  TMP A";
   %end;
%mend delfile;
```

Here is a call to the macro DELFILE in a PC environment that deletes a file named C:\SAS\SASUSER\DOC1.TMP:

```
%delfile(c:\sas\sasuser\doc1)
```

In this program, note the use of the portable %SYSEXEC statement to carry out the host-specific operating system commands.

Now, suppose you know your macro application is going to run on some version of Microsoft Windows. The SYSSCPL automatic macro variable provides information about the name of the host environment, similar to the SYSSCP automatic macro variable. However, SYSSCPL provides more information and enables you to further tailor your macro code.

Example Using SYSPARM

Suppose the SYSPARM= system option is set to the name of a city. That means the SYSPARM automatic variable is set to the name of that city. You can use that value to subset a data set and generate code specific to that value. Simply by making a small change to the command that invokes SAS (or to the configuration SAS file), your SAS job will perform different tasks.

```
/* Create a data set, based on the value of the */
/* SYSPARM automatic variable. */
/* An example data set name could be MYLIB.BOSTON. */
data mylib.&sysparm;
   set mylib.alltowns;
      /* Use the SYSPARM SAS language function to */
      /* compare the value (city name) */
      /* of SYSPARM to a data set variable. */
   if town=sysparm();
run;
```

When this program executes, you end up with a data set that contains data for only the town that you are interested in, and you can change what data set is generated before you start your SAS job.

Now suppose you want to further use the value of SYSPARM to control what procedures your job uses. The following macro does just that:

```
%macro select;
   %if %upcase(&sysparm) eq BOSTON %then
      %do;
         proc report ... more SAS code;
            title "Report on &sysparm";
         run;
      %end;

   %if %upcase(&sysparm) eq CHICAGO %then
      %do;
         proc chart ... more SAS code;
            title "Growth Values for &sysparm";
         run;
      %end;
   .
   .  /* more macro code */
   .
%mend select;
```

SYSPARM Details

The value of the SYSPARM automatic macro variable is the same as the value of the SYSPARM= system option, which is equivalent to the return value of the SAS language function SYSPARM. The default value is null. Because you can use the SYSPARM= system option at SAS invocation, you can set the value of the SYSPARM automatic macro variable before your SAS session begins.

SYSRC Details

The value of the SYSRC automatic macro variable contains the last return code generated by your host environment. The code returned is based on commands that you execute using the X statement in open code, the X command a windowing environment, or the %SYSEXEC macro statement (as well as the nonportable %TSO and %CMS macro statements). Use the SYSRC automatic macro variable to test the success or failure of a host environment command.

Note: While it does not generate an error message in the SAS log, the SYSRC automatic macro variable is not useful under all host environments. For example, under some host environments, the value of this variable is always 99, regardless of the success or failure of the host environment command. Check the SAS companion for your host environment to determine whether the SYSRC automatic macro variable is useful for your host environment.

Macro Language Elements with System Dependencies

Several macro language elements are host-specific, including the following:

any language element that relies on the sort sequence
: Examples of such expressions include %DO, %DO %UNTIL, %DO %WHILE, %IF-%THEN, and %EVAL.

For example, consider the following program:

```
%macro testsort(var);
    %if &var < a %then %put *** &var is less than a ***;
    %else %put *** &var is greater than a ***;
%mend testsort;
%testsort(1)
      /* Invoke the macro with the number 1 as the parameter. */
```

On EBCDIC systems, such as z/OS, and VSE, this program causes the following to be written to the SAS log:

```
*** 1 is greater than a ***
```

But on ASCII systems (such as UNIX or Windows), the following is written to the SAS log:

```
*** 1 is less than a ***
```

MSYMTABMAX=
: The MSYMTABMAX system option specifies the maximum amount of memory available to the macro variable symbol tables. If this value is exceeded, the symbol tables are stored in a WORK file on disk.

MVARSIZE=

The MVARSIZE system option specifies the maximum number of bytes for any macro variable stored in memory. If this value is exceeded, the macro variable is stored in a WORK file on disk.

%SCAN and %QSCAN

The default delimiters that the %SCAN and %QSCAN functions use to search for words in a string are different on ASCII and EBCDIC systems. The default delimiters are

ASCII systems

blank . < (+ & ! $ *) ; ^ − / , % |

EBCDIC systems

blank . < (+ | & ! $ *) ; ¬ − / , % ¦ ¢

%SYSEXEC, %TSO, and %CMS

The %SYSEXEC, %TSO, and %CMS macro statements enable you to issue a host environment command.

%SYSGET

On some host environments, the %SYSGET function returns the value of host environment variables and symbols.

SYSPARM=

The SYSPARM= system option can supply a value for the SYSPARM automatic macro variable at SAS invocation. It is useful in customizing a production job. For example, to create a title based on a city as part of noninteractive execution, the production program might contain the SYSPARM= system option in the SAS configuration file or the command that invokes SAS. See “SYSPARM Details” on page 152 for an example using the SYSPARM= system option in conjunction with the SYSPARM automatic macro variable.

SASMSTORE=

The SASMSTORE= system option specifies the location of stored compiled macros.

SASAUTOS=

The SASAUTOS= system option specifies the location of autocall macros.

Host-Specific Macro Variables

Some host environments create unique macro variables. These macro variables are not automatic macro variables. The following tables list some commonly used host-specific macro variables. Additional host-specific macro variables might be available in future releases. See your SAS companion for more details.

Table 11.3 *Host-Specific Macro Variables for z/OS*

Variable Name	Description
SYS99ERR	SVC99 error reason code
SYS99INF	SVC99 info reason code
SYS99MSG	YSC99 text message corresponding to the SVC error or info reason code
SYS99R15	SVC99 return code

Variable Name	Description
SYSJCTID	Value of the JCTUSER field in the JCT control block
SYSJMRID	Value of the JMRUSEID field in the JCT control block
SYSUID	TSO user ID associated with the SAS session

Naming Macros and External Files for Use with the Autocall Facility

When naming macros that will be stored in an autocall library, there are restrictions depending on your host environment. Here is a list of some of the restrictions:

- Every host environment has file naming conventions. If the host environment uses file extensions, use **`.sas`** as the extension of your macro files.
- Although SAS names can contain underscores, some host environments do not use them in the names of external files. Some host environments that do not use underscores do use the pound sign (#) and might automatically replace the # with _ when the macro is used.
- Some host environments have reserved words, such as CON and NULL. Do not use reserved words when naming autocall macros or external files.
- Some hosts have host-specific autocall macros. Do not define a macro with the same name as these autocall macros.
- Macro catalogs are not portable. Remember to always save your macro source code in a safe place.
- On UNIX systems the filename that contains the autocall macro must be all lowercase letters.

Chapter 12
Macro Language Elements

Macro Language Elements

The SAS macro language consists of statements, functions, and automatic macro variables. This section defines and lists these elements.

- “Macro Statements ” on page 156
- “Macro Functions ” on page 158
- “Automatic Macro Variables” on page 164

Also covered are the interfaces to the macro facility provided by Base SAS software, the SQL procedure, and SAS Component Language as well as selected autocall macros and macro system options.

Macro Statements

Using Macro Statements

A macro language statement instructs the macro processor to perform an operation. It consists of a string of keywords, SAS names, and special characters and operators, and it ends in a semicolon. Some macro language statements are used only in macro definitions, but you can use others anywhere in a SAS session or job, either inside or outside macro definitions (referred to as open code). The following table lists macro language statements that you can use in both macro definitions and open code.

Table 12.1 *Macro Language Statements Used in Macro Definitions and Open Code*

Statement	Description
%* comment	Designates comment text.
%COPY	Copies specified items from a SAS library.
%DISPLAY	Displays a macro window.
%GLOBAL	Creates macro variables that are available during the execution of an entire SAS session.
%INPUT	Supplies values to macro variables during macro execution.
%LET	Creates a macro variable and assigns it a value.
%MACRO	Begins a macro definition.
%PUT	Writes text or the values of macro variables to the SAS log.
%SYMDEL	Deletes the indicated macro variable named in the argument.
%SYSCALL	Invokes a SAS call routine.
%SYSEXEC	Issues operating system commands.
%SYSLPUT	Defines a new macro variable or modifies the value of an existing macro variable on a remote host or server.
%SYSMACDELETE	Deletes a macro definition from the WORK.SASMACR catalog.
%SYSMSTORECLEAR	Closes stored compiled macros and clears the SASMSTORE= library.
%SYSRPUT	Assigns the value of a macro variable on a remote host to a macro variable on the local host.
%WINDOW	Defines customized windows.

The following table lists macro language statements that you can use only in macro definitions.

Table 12.2 *Macro Language Statements Used in Macro Definitions Only*

Statement	Description
%ABORT	Stops the macro that is executing along with the current DATA step, SAS job, or SAS session.
%DO	Begins a %DO group.
%DO, Iterative	Executes statements repetitively, based on the value of an index variable.
%DO %UNTIL	Executes statements repetitively until a condition is true.
%DO %WHILE	Executes statements repetitively while a condition is true.
%END	Ends a %DO group.
%GOTO	Branches macro processing to the specified label.
%IF-%THEN/%ELSE	Conditionally processes a portion of a macro.
%label:	Identifies the destination of a %GOTO statement.
%LOCAL	Creates macro variables that are available only during the execution of the macro where they are defined.
%MEND	Ends a macro definition.
%RETURN	Causes normal termination of the currently executing macro.

Macro Statements That Perform Automatic Evaluation

Some macro statements perform an operation based on an evaluation of an arithmetic or logical expression. They perform the evaluation by automatically calling the %EVAL function. If you get an error message about a problem with %EVAL when a macro does not use %EVAL only, check for one of these statements. The following macro statements perform automatic evaluation:

- %DO *macro-variable–expression* %TO *expression* <%BY *expression*>;
- %DO %UNTIL(*expression*);
- %DO %WHILE(*expression*);
- %IF *expression* %THEN *action*;

For details about operands and operators in expressions, see "Macro Expressions" on page 71.

Macro Functions

Using Macro Functions

A macro language function processes one or more arguments and produces a result. You can use all macro functions in both macro definitions and open code. Macro functions include character functions, evaluation functions, and quoting functions. The macro language functions are listed in the following table.

Table 12.3 *Macro Functions*

Function	Description
%BQUOTE, %NRBQUOTE	Mask special characters and mnemonic operators in a resolved value at macro execution.
%EVAL	Evaluates arithmetic and logical expressions using integer arithmetic.
%INDEX	Returns the position of the first character of a string.
%LENGTH	Returns the length of a string.
%QUOTE, %NRQUOTE	Mask special characters and mnemonic operators in a resolved value at macro execution. Unmatched quotation marks (“ ”) and parentheses (()) must be marked with a preceding %.
%SCAN, %QSCAN	Search for a word specified by its number. %QSCAN masks special characters and mnemonic operators in its result.
%STR, %NRSTR	Mask special characters and mnemonic operators in constant text at macro compilation. Unmatched quotation marks (“ ”) and parentheses (()) must be marked with a preceding %.
%SUBSTR, %QSUBSTR	Produce a substring of a character string. %QSUBSTR masks special characters and mnemonic operators in its result.
%SUPERQ	Masks all special characters and mnemonic operators at macro execution but prevents resolution of the value.
%SYMEXIST	Returns an indication as to whether the named macro variable exists.
%SYMGLOBL	Returns an indication as to whether the named macro variable is global in scope.
%SYMLOCAL	Returns an indication as to whether the named macro variable is local in scope.
%SYSEVALF	Evaluates arithmetic and logical expressions using floating-point arithmetic.

Function	Description
%SYSFUNC, %QSYSFUNC	Execute SAS functions or user-written functions. %QSYSFUNC masks special characters and mnemonic operators in its result.
%SYSGET	Returns the value of a specified host environment variable.
%SYSMACEXEC	Indicates whether a macro is currently executing.
%SYSMACEXIST	Indicates whether there is a macro definition in the WORK.SASMACR catalog.
%SYSMEXECDEPTH	Returns the depth of nesting from the point of call.
%SYSMEXECNAME	Returns the name of the macro executing at a nesting level.
%SYSPROD	Reports whether a SAS software product is licensed at the site.
%UNQUOTE	Unmasks all special characters and mnemonic operators for a value.
%UPCASE, %QUPCASE	Convert characters to uppercase. %QUPCASE masks special characters and mnemonic operators in its result.

Macro Character Functions

Character functions change character strings or provide information about them. The following table lists the macro character functions.

Table 12.4 Macro Character Functions

Function	Description
%INDEX	Returns the position of the first character of a string.
%LENGTH	Returns the length of a string.
%SCAN, %QSCAN	Search for a word that is specified by a number. %QSCAN masks special characters and mnemonic operators in its result.
%SUBSTR, %QSUBSTR	Produce a substring of a character string. %QSUBSTR masks special characters and mnemonic operators in its result.
%UPCASE, %QUPCASE	Convert characters to uppercase. %QUPCASE masks special characters and mnemonic operators in its result.

For macro character functions that have a Q form (for example, %SCAN and %QSCAN), the two functions work alike except that the function beginning with Q masks special characters and mnemonic operators in its result. Use the function beginning with Q when an argument has been previously masked with a macro quoting function or when you want the result to be masked (for example, when the result might contain an unmatched quotation mark or parenthesis). For details, see “Macro Quoting” on page 80.

Many macro character functions have names corresponding to SAS character functions and perform similar tasks (such as %SUBSTR and SUBSTR). But, macro functions operate before the DATA step executes. Consider the following DATA step:

```
data out.%substr(&sysday,1,3);       /* macro function */
   set in.weekly (keep=name code sales);
   length location $4;
   location=substr(code,1,4);          /* SAS function */
run;
```

Running the program on Monday creates the data set name OUT.MON:

```
data out.MON;                        /* macro function */
   set in.weekly (keep=name code sales);
   length location $4;
   location=substr(code,1,4);          /* SAS function */
run;
```

Suppose that the IN.WEEKLY variable CODE contains the values cary18593 and apex19624. The SAS function SUBSTR operates during DATA step execution and assigns these values to the variable LOCATION: **`cary`** and **`apex`**.

Macro Evaluation Functions

Evaluation functions evaluate arithmetic and logical expressions. They temporarily convert the operands in the argument to numeric values. Then, they perform the operation specified by the operand and convert the result to a character value. The macro processor uses evaluation functions to do the following:

- make character comparisons
- evaluate logical (Boolean) expressions
- assign numeric properties to a token, such as an integer in the argument of a function

For more information, see “Macro Expressions” on page 71. The following table lists the macro evaluation functions.

Table 12.5 *Macro Evaluation Functions*

Function	Description
%EVAL	Evaluates arithmetic and logical expressions using integer arithmetic.
%SYSEVALF	Evaluates arithmetic and logical expressions using floating-point arithmetic.

%EVAL is called automatically by the macro processor to evaluate expressions in the arguments to the statements that perform evaluation in the following functions:

- %QSCAN(*argument, n<, delimiters>*)
- %QSUBSTR(*argument, position<, length>*)
- %SCAN(*argument, n<, delimiters>*)
- %SUBSTR(*argument, position<, length>*)

Macro Quoting Functions

Macro quoting functions mask special characters and mnemonic operators so the macro processor interprets them as text instead of elements of the macro language.

The following table lists the macro quoting functions, and also describes the special characters they mask and when they operate. (Although %QSCAN, %QSUBSTR, and %QUPCASE mask special characters and mnemonic operations in their results, they are not considered quoting functions because their purpose is to process a character value and not simply to quote a value.) For more information, see "Macro Quoting" on page 80.

Table 12.6 *Macro Quoting Functions*

Function	Description
%BQUOTE, %NRBQUOTE	Mask special characters and mnemonic operators in a resolved value at macro execution. %BQUOTE and %NRBQUOTE are the most powerful functions for masking values at execution time because they do not require that unmatched quotation marks (" ") and parentheses (()) be marked.
%QUOTE, %NRQUOTE	Mask special characters and mnemonic operators in a resolved value at macro execution. Unmatched quotation marks (" ") and parentheses (()) must be marked with a preceding %.
%STR, %NRSTR	Mask special characters and mnemonic operators in constant text at macro compilation. Unmatched quotation marks (" ") and parentheses (()) must be marked with a preceding %.
%SUPERQ	Masks all special characters and mnemonic operators at macro execution but prevents resolution of the value.
%UNQUOTE	Unmasks all special characters and mnemonic operators for a value.

Compilation Quoting Functions

%STR and %NRSTR mask special characters and mnemonic operators in values during compilation of a macro definition or a macro language statement in open code. For example, the %STR function prevents the following %LET statement from ending prematurely. It keeps the semicolon in the PROC PRINT statement from being interpreted as the semicolon for the %LET statement.

```
%let printit=%str(proc print; run;);
```

Execution of Macro Quoting Functions

%BQUOTE, %NRBQUOTE, %QUOTE, %NRQUOTE, and %SUPERQ mask special characters and mnemonic operators in values during execution of a macro or a macro language statement in open code. Except for %SUPERQ, these functions instruct the macro processor to resolve a macro expression as far as possible and mask the result, issuing warning messages for any macro variable references or macro invocations they cannot resolve. %SUPERQ protects the value of a macro variable from any attempt at further resolution.

Of the quoting functions that resolve values during execution, %BQUOTE and %NRBQUOTE are the most flexible. For example, the %BQUOTE function prevents the following %IF statement from producing an error if the macro variable STATE resolves to OR (for Oregon). Without %BQUOTE, the macro processor would interpret the abbreviation for Oregon as the logical operator OR.

```
%if %bquote(&state)=nc %then %put North Carolina Dept. of
Revenue;
```

%SUPERQ fetches the value of a macro variable from the macro symbol table and masks it immediately, preventing the macro processor from attempting to resolve any part of the resolved value. For example, %SUPERQ prevents the following %LET statement from producing an error when it resolves to a value with an ampersand, like **Smith&Jones**. Without %SUPERQ, the macro processor would attempt to resolve **&Jones**.

```
%let testvar=%superq(corpname);
     /* No ampersand in argument to %superq. */
```

(%SUPERQ takes as its argument either a macro variable name without an ampersand or a text expression that yields a macro variable name.)

Quotation Marks and Parentheses without a Match

Syntax errors result if the arguments of %STR, %NRSTR, %QUOTE, and %NRQUOTE contain a quotation mark or parenthesis that does not have a match. To prevent these errors, mark these quotation marks and parentheses by preceding them with a percent sign. For example, write the following to store the value **345)** in macro variable B:

```
%let b=%str(345%));
```

If an argument of %STR, %NRSTR, %QUOTE, or %NRQUOTE contains a percent sign that precedes a quotation mark or parenthesis, use two percent signs (%%) to specify that the argument's percent sign does not mark the quotation mark or parenthesis. For example, Write the following to store the value **TITLE "20%";** in macro variable P:

```
%let p=%str(TITLE "20%%";);
```

If the argument for one of these functions contains a character string with the comment symbols **/*** and **-->**, use a %STR function with each character. For example, consider these statements:

```
%let instruct=Comments can start with %str(/)%str(*).;
%put &instruct;
```

They write the following line to the SAS log:

```
Comments can start with /*
```

Note: Unexpected results can occur if the comment symbols are not quoted with a quoting function.

For more information about macro quoting, see "Macro Quoting" on page 80.

Macro Functions for Double-Byte Character Set (DBCS)

Because East Asian languages have thousands of characters, double (two) bytes of information are needed to represent each character. Each East Asian language usually has more than one DBCS encoding system. SAS processes the DBCS encoding information that is unique for the major East Asian languages. The following table defines the macro functions that support DBCS.

Table 12.7 *Macro Functions for DBCS*

Functions	Description
%KCMPRES	Compresses multiple blanks and removes leading and trailing blanks.
%KINDEX	Returns the position of the first character of a string.
%KLEFT and %QKLEFT	Left-aligns an argument by removing leading blanks.
%KLENGTH	Returns the length of a string.
%KSCAN and %QKSCAN	Searches for a word that is specified by its position in a string.
%KSUBSTR and %QKSUBSTR	%KSUBSTR and %QKSUBSTR produces a substring of a character string.
%KUPCASE and %QKUPCASE	Converts values to uppercase.

For more information, see "Macro Functions for NLS" in *SAS National Language Support (NLS): Reference Guide*.

Other Macro Functions

Seven other macro functions do not fit into the earlier categories, but they provide important information. The following table lists these functions.

Table 12.8 *Other Macro Functions*

Function	Description
%SYMEXIST	Returns an indication as to whether the named macro variable exists.
%SYMGLOBL	Returns an indication as to whether the named macro variable is global in scope.
%SYMLOCAL	Returns an indication as to whether the named macro variable is local in scope.

Function	Description
%SYSFUNC, %QSYSFUNC	Execute SAS language functions or user-written functions within the macro facility.
%SYSGET	Returns the value of the specified host environment variable. For details, see the SAS Companion for your operating environment.
%SYSPROD	Reports whether a SAS software product is licensed at the site.

The %SYSFUNC and %QSYSFUNC functions enable most of the functions from Base SAS software, a function written with the SAS/TOOLKIT software, or a function created using the FCMP procedure available to the macro facility. Consider the following examples:

- ```
  /* in a DATA step or SCL program */
     dsid=open("sasuser.houses","i");
  ```
- ```
  /* in the macro facility */
     %let dsid = %sysfunc(open(sasuser.houses,i));
  ```

For more information, see "%SYSFUNC and %QSYSFUNC Functions" on page 268.

Automatic Macro Variables

Automatic macro variables are created by the macro processor and they supply a variety of information. They are useful in programs to check the status of a condition before executing code. When you use automatic macro variables, you reference them the same way that you do macro variables that you create such as &SYSLAST or &SYSJOBID.

CAUTION:

Do not create macro variable names that begin with SYS. The three-letter prefix SYS is reserved for use by SAS for automatic macro variables. For a complete list of reserved words in the macro language, see " Reserved Words in the Macro Facility" on page 363.

For example, suppose you want to include the day and date that your current SAS session was invoked. Write the FOOTNOTE statement to reference the automatic macro variables SYSDAY and SYSDATE9:

```
footnote "Report for &sysday, &sysdate9";
```

If the current SAS session was invoked on June 13, 2007, macro variable resolution causes SAS to see this statement:

```
FOOTNOTE "Report for Friday, 13JUN2007";
```

All automatic variables except for SYSPBUFF are global and are created when you invoke SAS. The following table lists the automatic macro variables and describes their READ and WRITE status.

Table 12.9 Automatic Macro Variables

Variable	Read and Write Status
SYSADDRBITS	Read-only
SYSBUFFR	Read and write
SYSCC	Read and write
SYSCHARWIDTH	Read-only
SYSCMD	Read and write
SYSDATE	Read-only
SYSDATE9	Read-only
SYSDAY	Read-only
SYSDEVIC	Read and write
SYSDMG	Read and write
SYSDSN	Read and write
SYSENCODING	Read-only
SYSENDIAN	Read-only
SYSENV	Read-only
SYSERR	Read-only
SYSERRORTEXT	Read-only
SYSFILRC	Read and write
SYSHOSTNAME	Read-only
SYSINDEX	Read-only
SYSINFO	Read-only
SYSJOBID	Read-only
SYSLAST	Read and write
SYSLCKRC	Read and write
SYSLIBRC	Read and write
SYSLOGAPPLNAME	Read-only

Variable	Read and Write Status
SYSMACRONAME	Read-only
SYSMENV	Read-only
SYSMSG	Read and write
SYSNCPU	Read-only
SYSNOBS	Read-only
SYSODSESCAPECHAR	Read-only
SYSODSPATH	Read-only
SYSPARM	Read and write
SYSPBUFF	Read and write
SYSPROCESSID	Read-only
SYSPROCESSNAME	Read-only
SYSPROCNAME	Read-only
SYSRC	Read and write
SYSSCP	Read-only
SYSSCPL	Read-only
SYSSITE	Read-only
SYSSIZEOFLONG	Read-only
SYSSIZEOFPTR	Read-only
SYSSIZEOFUNICODE	Read-only
SYSSTARTID	Read-only
SYSSTARTNAME	Read-only
SYSTCPIPHOSTNAME	Read-only
SYSTIME	Read-only
SYSUSERID	Read-only
SYSVER	Read-only
SYSVLONG	Read-only

Variable	Read and Write Status
SYSVLONG4	Read-only
SYSWARNINGTEXT	Read-only

Interfaces with the Macro Facility

The DATA step, the SAS Component Language, and the SQL procedure provide interfaces with the macro facility. The following tables list the elements that interact with the SAS macro facility.

The DATA step provides elements that enable a program to interact with the macro facility during DATA step execution.

Table 12.10 *Interfaces to the DATA Steps*

Element	Description
EXECUTE routine	Resolves an argument and executes the resolved value at the next step boundary.
RESOLVE function	Resolves the value of a text expression during DATA step execution.
SYMDEL routine	Deletes the indicated macro variable named in the argument.
SYMEXIST function	Returns an indication as to whether the named macro variable exists.
SYMGET function	Returns the value of a macro variable to the DATA step during DATA step execution.
SYMGLOBL function	Returns an indication as to whether the named macro variable is global in scope.
SYMLOCAL function	Returns an indication as to whether the named macro variable is local in scope.
SYMPUT/SYMPUTX routines	Assigns a value produced in a DATA step to a macro variable.

The SAS Component Language (SCL) provides two elements for using the SAS macro facility to define macros and macro variables for SCL programs.

Table 12.11 *Interfaces to the SAS Component Language*

Element	Description
SYMGETN	Returns the value of a global macro variable as a numeric value.

Element	Description
SYMPUTN	Assigns a numeric value to a global macro variable.

The SQL procedure provides a feature for creating and updating macro variables with values produced by the SQL procedure.

Table 12.12 *Interfaces to the SQL Procedure*

Element	Description
INTO	Assigns the result of a calculation or the value of a data column.

For more information, see "Interfaces with the Macro Facility" on page 101.

Selected Autocall Macros Provided with SAS Software

Overview of Provided Autocall Macros

SAS supplies libraries of autocall macros to each SAS site. The libraries that you receive depend on the SAS products licensed at your site. You can use autocall macros without having to define or include them in your programs.

When SAS is installed, the autocall libraries are included in the value of the SASAUTOS system option in the system configuration file. The autocall macros are stored as individual members, each containing a macro definition. Each member has the same name as the macro definition that it contains.

Although the macros available in the autocall libraries supplied by SAS are working utility programs, you can also use them as models for your own routines. In addition, you can call them in macros that you write yourself.

To explore these macro definitions, browse the commented section at the beginning of each member. See the setting of SAS system option SASAUTOS, to find the location of the autocall libraries. To view the SASAUTOS value, use one of the following:

- the OPTIONS command in the SAS windowing environment to open the OPTIONS window
- the OPTIONS procedure
- the VERBOSE system option
- the OPLIST system option

For details about these options, see "SAS System Options," in *SAS System Options: Reference*.

The following table lists selected autocall macros.

Table 12.13 *Selected Autocall Macros*

Macro	Description
CMPRES and QCMPRES	Compresses multiple blanks and removes leading and trailing blanks. QCMPRES masks the result so special characters and mnemonic operators are treated as text instead of being interpreted by the macro facility.
COMPSTOR	Compiles macros and stores them in a catalog in a permanent SAS library.
DATATYP	Returns the data type of a value.
LEFT and QLEFT	Left-aligns an argument by removing leading blanks. QLEFT masks the result so special characters and mnemonic operators are treated as text instead of being interpreted by the macro facility.
SYSRC	Returns a value corresponding to an error condition.
TRIM and QTRIM	Trims trailing blanks. QTRIM masks the result so special characters and mnemonic operators are treated as text instead of being interpreted by the macro facility.
VERIFY	Returns the position of the first character unique to an expression.

Required System Options for Autocall Macros

To use autocall macros, you must set two SAS system options:

MAUTOSOURCE
: activates the autocall facility. NOMAUTOSOURCE disables the autocall facility.

SASAUTOS=*library-specification* | (*library-specification-1…, library-specification-n*)
: specifies the autocall library or libraries. For more information, see the SAS companion for your operating system.

If your site has installed the autocall libraries supplied by SAS and uses the standard configuration of SAS software supplied by SAS, you need only to ensure that the SAS system option MAUTOSOURCE is in effect to begin using the autocall macros.

Although the MAUTOLOCDISPLAY system option is not required, it displays the source location of the autocall macros in the SAS log when the autocall macro is invoked. For more information, see “MAUTOLOCDISPLAY System Option” on page 333.

Using Autocall Macros

To use an autocall macro, call it in your program with the statement %*macro-name*. The macro processor searches first in the WORK library for a compiled macro definition with that name. If the macro processor does not find a compiled macro and if the MAUTOSOURCE is in effect, the macro processor searches the libraries specified by the SASAUTOS option for a member with that name. When the macro processor finds the member, it does the following:

1. compiles all of the source statements in that member, including all macro definitions

2. executes any open code (macro statements or SAS source statements not within any macro definition) in that member
3. executes the macro with the name that you invoked

After the macro is compiled, it is stored in the WORK.SASMACR catalog and is available for use in the SAS session without having to be recompiled.

You can also create your own autocall macros and store them in libraries for easy execution. For more information, see "Storing and Reusing Macros" on page 113.

Autocall Macros for Double-Byte Character Set (DBCS)

Because East Asian languages have thousands of characters, double (two) bytes of information are needed to represent each character. Each East Asian language usually has more than one DBCS encoding system. SAS processes the DBCS encoding information that is unique for the major East Asian languages. The following table contains definitions for the autocall macros that support DBCS.

Table 12.14 *Autocall Macros for DBCS*

Autocall Macros	Description
%KLOWCASE and %QKLOWCAS	Changes the uppercase characters to lowercase.
%KTRIM and %QKTRIM	Trims the trailing blanks.
%KVERIFY	Returns the position of the first character unique to an expression.

For more information, see "Autocall Macros for NLS" in *SAS National Language Support (NLS): Reference Guide.*

Selected System Options Used in the Macro Facility

The following table lists the SAS system options that apply to the macro facility.

Table 12.15 *System Options Used in the Macro Facility*

Option	Description
CMDMAC	Controls command-style macro invocation.
IMPLMAC	Controls statement-style macro invocation.
MACRO	Controls whether the SAS macro language is available.
MAUTOCOMPLOC	Displays in the SAS log the source location of the autocall macros when the autocall macro is compiled.

Option	Description
MAUTOLOCDISPLAY	Displays the source location of the autocall macros in the SAS log when the autocall macro is invoked.
MAUTOLOCINDES	Specifies whether the macro processor prepends the full pathname of the autocall source file to the description field of the catalog entry of compiled auto call macro definition in the WORK.SASMACR catalog.
MAUTOSOURCE	Controls whether the macro autocall feature is available.
MCOMPILE	Allows new definitions of macros.
MCOMPILENOTE	Issues a NOTE to the SAS log upon the completion of the compilation of a macro.
MCOVERAGE	Enables the generation of coverage analysis data.
MCOVERAGELOC	Specifies the location of the coverage analysis data file.
MERROR	Controls whether the macro processor issues a warning message when a macro-like name (*%name*) does not match a compiled macro.
MEXECNOTE	Displays macro execution information in the SAS log at macro invocation.
MEXECSIZE	Specifies the maximum macro size that can be executed in memory.
MFILE	Determines whether MPRINT output is routed to an external file.
MINDELIMITER	Specifies the character to be used as the delimiter for the macro IN operator.
MINOPERATOR	Controls whether the macro processor recognizes the IN (#) logical operator.
MLOGIC	Controls whether macro execution is traced for debugging.
MLOGICNEST	Allows the macro nesting information to be displayed in the MLOGIC output in the SAS log.
MPRINT	Controls whether SAS statements generated by macro execution are traced for debugging.
MPRINTNEST	Allows the macro nesting information to be displayed in the MPRINT output in the SAS log.
MRECALL	Controls whether the macro processor searches the autocall libraries for a member that was not found during an earlier search.
MREPLACE	Enables existing macros to be redefined.

Option	Description
MSTORED	Controls whether stored compiled macros are available.
MSYMTABMAX	Specifies the maximum amount of memory available to the macro variable symbol table or tables.
MVARSIZE	Specifies the maximum size for in-memory macro variable values.
SASAUTOS	Specifies one or more autocall libraries.
SASMSTORE	Specifies the libref of a SAS library containing a catalog of stored compiled SAS macros.
SERROR	Controls whether the macro processor issues a warning message when a macro variable reference does not match a macro variable.
SYMBOLGEN	Controls whether the results of resolving macro variable references are displayed for debugging.
SYSPARM	Specifies a character string that can be passed to SAS programs.

Part 2

Macro Language Dictionary

Chapter 13
AutoCall Macros

AutoCall Macros

SAS supplies libraries of autocall macros to each SAS site. The libraries that you receive depend on the SAS products licensed at your site. You can use autocall macros without having to define or include them in your programs.

Dictionary

%CMPRES and %QCMPRES Autocall Macros

Compress multiple blanks and remove leading and trailing blanks.

Type: Autocall macros

Requirement: MAUTOSOURCE system option

Syntax

%CMPRES (*text* | *text expression*)

%QCMPRES (*text* | *text expression*)

Details

Note: Autocall macros are included in a library supplied by SAS Institute. This library might not be installed at your site or might be a site-specific version. If you cannot access this macro or if you want to find out if it is a site-specific version, see your on-site SAS support personnel. For more information, see "Storing and Reusing Macros" on page 113.

The CMPRES and QCMPRES macros compress multiple blanks and remove leading and trailing blanks. If the argument might contain a special character or mnemonic operator, listed below, use %QCMPRES.

CMPRES returns an unquoted result, even if the argument is quoted. QCMPRES produces a result with the following special characters and mnemonic operators masked, so the macro processor interprets them as text instead of as elements of the macro language:

```
& % ' " ( ) + - * / < > = ¬ ^ ~ ; , # blank
AND OR NOT EQ NE LE LT GE GT IN
```

Examples

Example 1: Removing Unnecessary Blanks with %CMPRES

```
%macro createft;
   %let footnote="The result of &x &op &y is %eval(&x &op &y).";
   footnote1 &footnote;
   footnote2 %cmpres(&footnote);
%mend createft;
data _null_;
   x=5;
   y=10;
   call symput('x',x);     /* Uses BEST12. format */
   call symput('y',y);     /* Uses BEST12. format */
   call symput('op','+'); /* Uses $1. format      */
run;
%createft
```

The CREATEFT macro generates two footnote statements.

```
FOOTNOTE1 "The result of 5 + _________10 is _________15.";
FOOTNOTE2 "The result of 5 + 10 is 15.";
```

Example 2: Contrasting %QCMPRES and %CMPRES

```
%let x=5;
%let y=10;
%let a=%nrstr(%eval(&x   +   &y));
%put QCMPRES: %qcmpres(&a);
%put CMPRES: %cmpres(&a);
```

The %PUT statements write the following lines to the log:

```
QCMPRES: %eval(&x + &y)
CMPRES: 15
```

%COMPSTOR Autocall Macro

Compiles macros and stores them in a catalog in a permanent SAS library.

Type: Autocall macro

Requirement: MAUTOSOURCE system option

Syntax

%COMPSTOR (PATHNAME=*SAS library*)

Required Argument

SAS-data-library
is the physical name of a SAS library on your host system. The COMPSTOR macro uses this value to automatically assign a libref. Do not enclose *SAS library* in quotation marks.

Details

Note: Autocall macros are included in a library supplied by SAS. This library might not be installed at your site or might be a site-specific version. If you cannot access this macro or if you want to find out if it is a site-specific version, see your on-site SAS support personnel. For more information, see "Storing and Reusing Macros" on page 113.

The COMPSTOR macro compiles the following autocall macros in a SAS catalog named SASMACR in a permanent SAS library. The overhead of compiling is saved when these macros are called for the first time in a SAS session. You can use the COMPSTOR macro as an example of how to create compiled stored macros. For more information about the autocall macros that are supplied by SAS or about using stored compiled macros, see "Storing and Reusing Macros" on page 113.

%CMPRES	%QLEFT
%DATATYP	%QTRIM
%LEFT	%TRIM
%QCMPRES	%VERIFY

%DATATYP Autocall Macro

Returns the data type of a value.

Type: Autocall macro

Restriction: Autocall macros are included in a library supplied by SAS. This library might not be installed at your site or might be a site-specific version. If you cannot access this macro or if you want to find out if it is a site-specific version, see your on-site SAS support personnel.

Requirement: MAUTOSOURCE system option

Syntax

%DATATYP (*text* | *text expression*)

Details

The DATATYP macro returns a value of **NUMERIC** when an argument consists of digits and a leading plus or minus sign, a decimal, or a scientific or floating-point exponent (E or D in uppercase or lowercase letters). Otherwise, it returns the value **CHAR.**

Note: %DATATYP does not identify hexadecimal numbers.

Example: Determining the Data Type of a Value

```
%macro add(a,b);
%if (%datatyp(&a)=NUMERIC and %datatyp(&b)=NUMERIC) %then %do;
    %put The result is %sysevalf(&a+&b).;
%end;
%else %do;
   %put Error:  Addition requires numbers.;
%end;
%mend add;
```

You can invoke the ADD macro as:

```
%add(5.1E2,225)
```

The macro then writes this message to the SAS log:

```
The result is 735.
```

Similarly, you can invoke the ADD macro as:

```
%add(0c1x, 12)
```

The macro then writes this message to the SAS log:

```
Error:  Addition requires numbers.
```

%KVERIFY Autocall Macro

Returns the position of the first character unique to an expression.

Category: DBCS

Type: Autocall macro for NLS

Requirement: MAUTOSOURCE system option

Syntax

%KVERIFY(*source*, *excerpt*)

Required Arguments

source
: text or a text expression. This is the text that you want to examine for characters that do not exist in the excerpt.

excerpt
: text or a text expression. This is the text that defines the set of characters that %KVERIFY uses to examine the source.

Details

Note: Autocall macros are included in a library supplied by SAS. This library might not be installed at your site or might be a site-specific version. If you cannot access this macro or if you want to find out if it is a site-specific version, see your on-site SAS support personnel.

%KVERIFY returns the position of the first character in the source that is not also present in excerpt. If all the characters in source are present in the excerpt, %KVERIFY returns a value of 0.

%LEFT and %QLEFT Autocall Macro

Left-align an argument by removing leading blanks.

Type: Autocall macro

Requirement: MAUTOSOURCE system option

Syntax

%LEFT(*text* | *text expression*)

%QLEFT(*text* | *text expression*)

Details

Note: Autocall macros are included in a library supplied by SAS. This library might not be installed at your site or might be a site-specific version. If you cannot access this macro or if you want to find out if it is a site-specific version, see your on-site SAS support personnel. For more information, see "Storing and Reusing Macros" on page 113.

The LEFT macro and the QLEFT macro both left-align arguments by removing leading blanks. If the argument might contain a special character or mnemonic operator, listed below, use %QLEFT.

%LEFT returns an unquoted result, even if the argument is quoted. %QLEFT produces a result with the following special characters and mnemonic operators masked so the macro processor interprets them as text instead of as elements of the macro language:

```
& % ' " ( ) + - * / < > = ¬ ^ ~ ; , # blank
AND OR NOT EQ NE LE LT GE GT IN
```

Example: Contrasting %LEFT and %QLEFT

In this example, both the LEFT and QLEFT macros remove leading blanks. However, the QLEFT macro protects the leading & in the macro variable SYSDAY so it does not resolve.

```
%let d=%nrstr(   &sysday   );
%put *&d* *%qleft(&d)* *%left(&d)*;
```

The %PUT statement writes the following line to the SAS log:

```
*   &sysday    * *&sysday    * *Tuesday    *
```

%LOWCASE and %QLOWCASE Autocall Macros

Change uppercase characters to lowercase.

Type: Autocall macros

Requirement: MAUTOSOURCE system option

Syntax

%LOWCASE (*text* | *text expression*)

%QLOWCASE (*text* | *text expression*)

Details

Note: Autocall macros are included in a library supplied by SAS. This library might not be installed at your site or might be a site-specific version. If you cannot access this macro or if you want to find out if it is a site-specific version, see your on-site SAS support personnel. For more information, see “Storing and Reusing Macros” on page 113.

The %LOWCASE and %QLOWCASE macros change uppercase alphabetic characters to their lowercase equivalents. If the argument might contain a special character or mnemonic operator, listed below, use %QLOWCASE.

%LOWCASE returns a result without quotation marks, even if the argument has quotation marks. %QLOWCASE produces a result with the following special characters and mnemonic operators masked so the macro processor interprets them as text instead of as elements of the macro language:

```
& % ' " ( ) + - * / < > = ¬ ^ ~ ; , # blank
AND OR NOT EQ NE LE LT GE GT IN
```

Example: Creating a Title with Initial Letters Capitalized

```
%macro initcaps(title);
   %global newtitle;
   %let newtitle=;
   %let lastchar=;
   %do i=1 %to %length(&title);
      %let char=%qsubstr(&title,&i,1);
      %if (&lastchar=%str( ) or &i=1) %then %let char=%qupcase(&char);
      %else %let char=%qlowcase(&char);
```

```
      %let newtitle=&newtitle&char;
      %let lastchar=&char;
   %end;
   TITLE "&newtitle";
%mend;
%initcaps(%str(sales: COMMAND REFERENCE, VERSION 2, SECOND EDITION))
```

Submitting this example generates the following statement:

```
TITLE "Sales: Command Reference, Version 2, Second Edition";
```

%QCMPRES Autocall Macro

Compresses multiple blanks, removes leading and trailing blanks, and returns a result that masks special characters and mnemonic operators.

Type: Autocall macro

Requirement: MAUTOSOURCE system option

Syntax

%QCMPRES (*text* | *text expression*)

Without Arguments

See "%CMPRES and %QCMPRES Autocall Macros" on page 175.

Details

Note: Autocall macros are included in a library supplied by SAS. This library might not be installed at your site or might be a site-specific version. If you cannot access this macro or if you want to find out if it is a site-specific version, see your on-site SAS support personnel. For more information, see "Storing and Reusing Macros" on page 113.

%QLEFT Autocall Macro

Left-aligns an argument by removing leading blanks and returns a result that masks special characters and mnemonic operators.

Type: Autocall macro

Requirement: MAUTOSOURCE system option

Syntax

%QLEFT (*text* | *text expression*)

Without Arguments

See "%LEFT and %QLEFT Autocall Macro" on page 179.

Details

Note: Autocall macros are included in a library supplied by SAS. This library might not be installed at your site or might be a site-specific version. If you cannot access this macro or if you want to find out if it is a site-specific version, see your on-site SAS support personnel. For more information, see "Storing and Reusing Macros" on page 113.

%QLOWCASE Autocall Macro

Changes uppercase characters to lowercase and returns a result that masks special characters and mnemonic operators.

Type: Autocall macro

Requirement: MAUTOSOURCE system option

Syntax

%QLOWCASE(*text* | *text expression*)

Without Arguments

See "%LOWCASE and %QLOWCASE Autocall Macros" on page 180.

Details

Note: Autocall macros are included in a library supplied by SAS. This library might not be installed at your site or might be a site-specific version. If you cannot access this macro or if you want to find out if it is a site-specific version, see your on-site SAS support personnel. For more information, see "Storing and Reusing Macros" on page 113.

%QTRIM Autocall Macro

Trims trailing blanks and returns a result that masks special characters and mnemonic operators.

Type: Autocall macro

Requirement: MAUTOSOURCE system option

Syntax

%QTRIM (*text* | *text expression*)

Without Arguments

See "%TRIM and %QTRIM Autocall Macro" on page 188.

Details

Note: Autocall macros are included in a library supplied by SAS. This library might not be installed at your site or might be a site-specific version. If you cannot access this macro or if you want to find out if it is a site-specific version, see your on-site SAS support personnel. For more information, see “Storing and Reusing Macros” on page 113.

%SYSRC Autocall Macro

Returns a value corresponding to an error condition.

Type: Autocall macro

Requirement: MAUTOSOURCE system option

Syntax

%SYSRC(*character-string*)

Required Argument

character-string
is one of the mnemonic values listed in Table 13.1 on page 183 or a text expression that produces the mnemonic value.

Details

Note: Autocall macros are included in a library supplied by SAS. This library might not be installed at your site or might be a site-specific version. If you cannot access this macro or if you want to find out if it is a site-specific version, see your on-site SAS support personnel. For more information, see “Storing and Reusing Macros” on page 113.

The SYSRC macro enables you to test for return codes produced by SCL functions, the MODIFY statement, and the SET statement with the KEY= option. The SYSRC autocall macro tests for the error conditions by using mnemonic strings rather than the numeric values associated with the error conditions.

When you invoke the SYSRC macro with a mnemonic string, the macro generates a SAS return code. The mnemonics are easier to read than the numeric values, which are not intuitive and subject to change.

You can test for specific errors in SCL functions by comparing the value returned by the function with the value returned by the SYSRC macro with the corresponding mnemonic. To test for errors in the most recent MODIFY or SET statement with the KEY= option, compare the value of the _IORC_ automatic variable with the value returned by the SYSRC macro when you invoke it with the value of the appropriate mnemonic.

The following table lists the mnemonic values to specify with the SYSRC function and a description of the corresponding error.

Table 13.1 *Mnemonics for Warning and Error Conditions*

Mnemonic	Description
Library Assign or Deassign Messages	
_SEDUPLB	The libref refers to the same physical library as another libref.

Mnemonic	Description
_SEIBASN	The specified libref is not assigned.
_SEINUSE	The library or member is not available for use.
_SEINVLB	The library is not in a valid format for the access method.
_SEINVLN	The libref is not valid.
_SELBACC	The action requested cannot be performed because you do not have the required access level on the library.
_SELBUSE	The library is still in use.
_SELGASN	The specified libref is not assigned.
_SENOASN	The libref is not assigned.
_SENOLNM	The libref is not available for use.
_SESEQLB	The library is in sequential (tape) format.
_SWDUPLB	The libref refers to the same physical file as another libref.
_SWNOLIB	The library does not exist.
Fileref Messages	
_SELOGNM	The fileref is assigned to an invalid file.
_SWLNASN	The fileref is not assigned.
SAS Data Set Messages	
_DSENMR	The TRANSACTION data set observation does not exist in the MASTER data set.
_DSEMTR	Multiple TRANSACTION data set observations do not exist in MASTER data set.
_DSENOM	No matching observation was found in MASTER data set.
_SEBAUTH	The data set has passwords.
_SEBDIND	The index name is not a valid SAS name.
_SEDSMOD	The data set is not open in the correct mode for the specified operation.
_SEDTLEN	The data length is invalid.
_SEINDCF	The new name conflicts with an index name.

Mnemonic	Description
_SEINVMD	The open mode is invalid.
_SEINVPN	The physical name is invalid.
_SEMBACC	You do not have the level of access required to open the data set in the requested mode.
_SENOLCK	A record-level lock is not available.
_SENOMAC	Member-level access to the data set is denied.
_SENOSAS	The file is not a SAS data set.
_SEVARCF	The new name conflicts with an existing variable name.
_SWBOF	You tried to read the previous observation when you were on the first observation.
_SWNOWHR	The record no longer satisfies the WHERE clause.
_SWSEQ	The task requires reading observations in a random order, but the engine that you are using allows only sequential access.
_SWWAUG	The WHERE clause has been augmented.
_SWWCLR	The WHERE clause has been cleared.
_SWWREP	The WHERE clause has been replaced.
SAS File Open and Update Messages	
_SEBDSNM	The filename is not a valid SAS name.
_SEDLREC	The record has been deleted from the file.
_SEFOPEN	The file is currently open.
_SEINVON	The option name is invalid.
_SEINVOV	The option value is invalid.
_SEINVPS	The value of the File Data Buffer pointer is invalid.
_SELOCK	The file is locked by another user.
_SENOACC	You do not have the level of access required to open the file in the requested mode.
_SENOALL	_ALL_ is not allowed as part of a filename in this release.
_SENOCHN	The record was not changed because it would cause a duplicate value for an index that does not allow duplicates.

Mnemonic	Description
_SENODEL	Records cannot be deleted from this file.
_SENODLT	The file could not be deleted.
_SENOERT	The file is not open for writing.
_SENOOAC	You are not authorized for the requested open mode.
_SENOOPN	The file or directory is not open.
_SENOPF	The physical file does not exist.
_SENORD	The file is not opened for reading.
_SENORDX	The file is not radix addressable.
_SENOTRD	No record has been read from the file yet.
_SENOUPD	The file cannot be opened for update because the engine is read only.
_SENOWRT	You do not have write access to the member.
_SEOBJLK	The file or directory is in exclusive use by another user.
_SERECRD	No records have been read from the input file.
_SWACMEM	Access to the directory will be provided one member at a time.
_SWDLREC	The record has been deleted from file.
_SWEOF	End of file.
_SWNOFLE	The file does not exist.
_SWNOPF	The file or directory does not exist.
_SWNOREP	The file was not replaced because of the NOREPLACE option.
_SWNOTFL	The item pointed to exists but is not a file.
_SWNOUPD	This record cannot be updated at this time.
Library/Member/Entry Messages	
_SEBDMT	The member type specification is invalid.
_SEDLT	The member was not deleted.
_SELKUSR	The library or library member is locked by another user.

Mnemonic	Description
_SEMLEN	The member name is too long for this system.
_SENOLKH	The library or library member is not currently locked.
_SENOMEM	The member does not exist.
_SWKNXL	You have locked a library, member, or entry, that does not exist yet.
_SWLKUSR	The library or library member is locked by another user.
_SWLKYOU	You have already locked the library or library member.
_SWNOLKH	The library or library member is not currently locked.
Miscellaneous Operations	
_SEDEVOF	The device is offline or unavailable.
_SEDSKFL	The disk or tape is full.
_SEINVDV	The device type is invalid.
_SENORNG	There is no write ring in the tape opened for write access.
_SOK	The function was successful.
_SWINVCC	The carriage-control character is invalid.
_SWNODSK	The device is not a disk.
_SWPAUAC	Pause in I/O, process accumulated data up to this point.
_SWPAUSL	Pause in I/O, slide data window forward and process accumulated data up to this point.
_SWPAUU1	Pause in I/O, extra user control point 1.
_SWPAUU2	Pause in I/O, extra user control point 2.

Comparisons

The SYSRC autocall macro and the SYSRC automatic macro variable are not the same. For more information, see “SYSRC Automatic Macro Variable” on page 212.

Example: Examining the Value of _IORC_

The following DATA step illustrates using the autocall macro SYSRC and the automatic variable _IORC_ to control writing a message to the SAS log:

```
data big;
   modify big trans;
   by id;
   if _iorc_=%sysrc(_dsenmr) then put 'WARNING: Check ID=' id;
run;
```

%TRIM and %QTRIM Autocall Macro

Trim trailing blanks.

Type: Autocall macro

Requirement: MAUTOSOURCE system option

Syntax

%TRIM(*text* | *text expression*)

%QTRIM(*text* | *text expression*)

Details

Note: Autocall macros are included in a library supplied by SAS. This library might not be installed at your site or might be a site-specific version. If you cannot access this macro or if you want to find out if it is a site-specific version, see your on-site SAS support personnel. For more information, see "Storing and Reusing Macros" on page 113.

The TRIM macro and the QTRIM macro both trim trailing blanks. If the argument contains a special character or mnemonic operator, listed below, use %QTRIM.

QTRIM produces a result with the following special characters and mnemonic operators masked so the macro processor interprets them as text instead of as elements of the macro language:

```
& % ' " ( ) + - * / < > = ¬ ° ~ ; , #  blank
AND OR NOT EQ NE LE LT GE GT IN
```

Examples

Example 1: Removing Trailing Blanks

In this example, the TRIM autocall macro removes the trailing blanks from a message that is written to the SAS log.

```
%macro numobs(dsn);
%local num;
data _null_;
   set &dsn nobs=count;
   call symput('num', left(put(count,8.)));
   stop;
   run;
    %if &num eq 0 %then
       %put There were NO observations in %upcase(&dsn).;
    %else
       %put There were %trim(&num) observations in %upcase(&dsn).;
```

```
%mend numobs;
%numobs(sample)
```

Invoking the NUMOBS macro generates the following statements:

```
DATA _NULL_;
SET SAMPLE NOBS=COUNT;
CALL SYMPUT('num', LEFT(PUT(COUNT,8.)));
STOP;
RUN;
```

If the data set SAMPLE contains six observations, then the %PUT statement writes this line to the SAS log:

```
There were 6 observations in SAMPLE.
```

Example 2: Contrasting %TRIM and %QTRIM

These statements are executed January 28, 1999:

```
%let date=%nrstr(   &sysdate   );
%put *&date* *%qtrim(&date)* *%trim(&date)*;
```

The %PUT statement writes this line to the SAS log:

```
*   &sysdate   * *   &sysdate* *   28JAN99*
```

%VERIFY Autocall Macro

Returns the position of the first character unique to an expression.

Type: Autocall macro

Requirement: MAUTOSOURCE system option

Syntax

%VERIFY(*source, excerpt*)

Required Arguments

source
is text or a text expression that you want to examine for characters that do not exist in *excerpt*.

excerpt
is text or a text expression. This is the text that defines the set of characters that %VERIFY uses to examine *source*.

Details

Note: Autocall macros are included in a library supplied by SAS. This library might not be installed at your site or might be a site-specific version. If you cannot access this macro or if you want to find out if it is a site-specific version, see your on-site SAS support personnel. For more information, see "Storing and Reusing Macros" on page 113.

%VERIFY returns the position of the first character in *source* that is not also present in *excerpt*. If all characters in *source* are present in *excerpt*, %VERIFY returns 0.

Example: Testing for a Valid Fileref

The ISNAME macro checks a string to verify that it is a valid fileref and prints a message in the SAS log that explains why a string is or is not valid.

```
%macro isname(name);
    %let name=%upcase(&name);
    %if %length(&name)>8 %then
       %put &name: The fileref must be 8 characters or less.;
    %else %do;
       %let first=ABCDEFGHIJKLMNOPQRSTUVWXYZ_;
       %let all=&first.1234567890;
       %let chk_1st=%verify(%substr(&name,1,1),&first);
       %let chk_rest=%verify(&name,&all);
       %if &chk_rest>0 %then
          %put &name: The fileref cannot contain
              "%substr(&name,&chk_rest,1)".;
       %if &chk_1st>0 %then
          %put &name: The first character cannot be
              "%substr(&name,1,1)".;
       %if (&chk_1st or &chk_rest)=0  %then
          %put &name is a valid fileref.;
    %end;
 %mend isname;
%isname(file1)
%isname(1file)
%isname(filename1)
%isname(file$)
```

When this program executes, the following is written to the SAS log:

```
FILE1 is a valid fileref.
1FILE: The first character cannot be "1".
FILENAME1: The fileref must be 8 characters or less.
FILE$: The fileref cannot contain "$".
```

Chapter 14
Automatic Macro Variables

Automatic Macro Variables

Automatic macro variables are created by the macro processor and they supply a variety of information. They are useful in programs to check the status of a condition before executing code.

Dictionary

SYSADDRBITS Automatic Macro Variable

Contains the number of bits of an address.

Type: Automatic macro variable (read only)

Details

The SYSADDRBITS automatic macro variable contains the number of bits needed for an address.

SYSBUFFR Automatic Macro Variable

Contains text that is entered in response to a %INPUT statement when there is no corresponding macro variable.

Type: Automatic macro variable (read and write)

Details

Until the first execution of a %INPUT statement, SYSBUFFR has a null value. However, SYSBUFFR receives a new value during each execution of a %INPUT statement, either the text entered in response to the %INPUT statement where there is no corresponding macro variable or a null value. If a %INPUT statement contains no macro variable names, all characters entered are assigned to SYSBUFFR.

Example: Assigning Text to SYSBUFFR

This %INPUT statement accepts the values of the two macro variables WATRFALL and RIVER:

```
%input watrfall river;
```

If you enter the following text, there is not a one-to-one match between the two variable names and the text:

```
Angel Tributary of Caroni
```

For example, you can submit these statements:

```
%put WATRFALL contains: *&watrfall*;
%put RIVER contains: *&river*;
%put SYSBUFFR contains: *&sysbuffr*;
```

After execution, they produce this output in the SAS log:

```
WATRFALL contains: *Angel*
RIVER contains: *Tributary*
SYSBUFFR contains: * of Caroni*
```

As the SAS log demonstrates, the text stored in SYSBUFFR includes leading and embedded blanks.

SYSCC Automatic Macro Variable

Contains the current condition code that SAS returns to your operating environment (the operating environment condition code).

Type: Automatic macro variable (read and write)

Details

SYSCC is a read and write automatic macro variable that enables you to reset the job condition code and to recover from conditions that prevent subsequent steps from running.

A normal exit internally to SAS is 0. The host code translates the internal value to a meaningful condition code by each host for each operating environment. &SYSCC of 0 at SAS termination is the value of success for that operating environment's return code.

The following are examples of successful condition codes:

Table 14.1 *SYSCC Operating Environments and Values*

Operating Environment	Value
z/OS	RC 0
OpenVMS	$STATUS = 1

The method to check the operating environment return code is host dependent.

The warning condition code in SAS sets &SYSCC to 4.

Note: When the ERRORCHECK= SAS system option is set at NORMAL, the value of SYSCC will be 0 even if an error exists in a LIBNAME or FILENAME statement, or in a LOCK statement in SAS/SHARE software. The value of SYSCC will also be 0 when the %INCLUDE statement fails due to a nonexistent file. For more information, see the "ERRORCHECK= System Option" in *SAS System Options: Reference*.

SYSCHARWIDTH Automatic Macro Variable

Contains the character width value.

Type: Automatic macro variable (read only)

Details

The character width value is either 1 (narrow) or 2 (wide).

SYSCMD Automatic Macro Variable

Contains the last unrecognized command from the command line of a macro window.

Type: Automatic macro variable (read and write)

Details

The value of SYSCMD is null before each execution of a %DISPLAY statement. If you enter a word or phrase on the command line of a macro window and the windowing environment does not recognize the command, SYSCMD receives that word or phrase as its value. This method is the only way to change the value of SYSCMD, which otherwise is a read-only variable. Use SYSCMD to enter values on the command line that work like user-created windowing commands.

Example: Processing Commands Entered in a Macro Window

The macro definition START creates a window in which you can use the command line to enter any windowing command. If you type an invalid command, a message informs you that the command is not recognized. When you type QUIT on the command line, the window closes and the macro terminates.

```
%macro start;
  %window start
     #5  @28 'Welcome to the SAS System'
     #10  @28 'Type QUIT to exit';
  %let exit = 0;
  %do %until (&exit=1);
     %display start;
     %if &syscmd ne %then %do;
```

```
         %if %upcase(&syscmd)=QUIT %then %let exit=1;
         %else %let sysmsg=&syscmd not recognized;
      %end;
   %end;
%mend start;
```

SYSDATE Automatic Macro Variable

Contains the date that a SAS job or session began executing.

Type: Automatic macro variable (read only)

See: "SYSDATE9 Automatic Macro Variable" on page 195

Details

SYSDATE contains a SAS date value in the DATE7. format, which displays a two-digit date, the first three letters of the month name, and a two-digit year. The date does not change during the individual job or session. For example, you could use SYSDATE in programs to check the date before you execute code that you want to run on certain dates of the month.

Example: Formatting a SYSDATE Value

Macro FDATE assigns a format that you specify to the value of SYSDATE:

```
%macro fdate(fmt);
   %global fdate;
   data _null_;
      call symput("fdate",left(put("&sysdate"d,&fmt)));
   run;
%mend fdate;
%fdate(worddate.)
title "Tests for &fdate";
```

If you execute this macro on July 28, 1998, SAS sees the statements:

```
DATA _NULL_;
   CALL SYMPUT("FDATE",LEFT(PUT("28JUL98"D,WORDDATE.)));
RUN;
TITLE "Tests for July 28, 1998";
```

For another method of formatting the current date, see the %SYSFUNC and %QSYSFUNC functions.

SYSDATE9 Automatic Macro Variable

Contains the date that a SAS job or session began executing.

Type: Automatic macro variable (read only)

See: "SYSDATE Automatic Macro Variable" on page 195

Details

SYSDATE9 contains a SAS date value in the DATE9. format, which displays a two-digit date, the first three letters of the month name, and a four-digit year. The date does not change during the individual job or session. For example, you could use SYSDATE9 in programs to check the date before you execute code that you want to run on certain dates of the month.

Example: Formatting a SYSDATE9 Value

Macro FDATE assigns a format that you specify to the value of SYSDATE9:

```
%macro fdate(fmt);
b    %global fdate;
   data _null_;
      call symput("fdate",left(put("&sysdate9"d,&fmt)));
   run;
%mend fdate;
%fdate(worddate.)
title "Tests for &fdate";
```

If you execute this macro on July 28, 2008, SAS sees the statements:

```
DATA _NULL_;
   CALL SYMPUT("FDATE",LEFT(PUT("28JUL2008"D,WORDDATE.)));
RUN;
TITLE "Tests for July 28, 2008";
```

For another method of formatting the current date, see the %SYSFUNC and %QSYSFUNC functions.

SYSDAY Automatic Macro Variable

Contains the day of the week that a SAS job or session began executing.

Type: Automatic macro variable (read only)

Details

You can use SYSDAY to check the current day before executing code that you want to run on certain days of the week, provided you initialized your SAS session today.

Example: Identifying the Day When a SAS Session Started

The following statement identifies the day and date when a SAS session started running.

```
%put This SAS session started running on: &sysday, &sysdate9.;
```

When this statement executes on Wednesday, December 19, 2007 for a SAS session that began executing on Monday, December 17, 2007, the following line is written to the SAS log:

```
This SAS session started running on: Monday, 17DEC2007
```

SYSDEVIC Automatic Macro Variable

Contains the name of the current graphics device.

Type: Automatic macro variable (read and write)

Details

The current graphics device is the one specified at invocation of SAS. You can specify the graphics device on the command line in response to a prompt when you use a product that uses SAS/GRAPH. You can also specify the graphics device in a configuration file. The name of the current graphics device is also the value of the SAS system option DEVICE=.

For details, see the SAS documentation for your operating environment.

Note: The macro processor always stores the value of SYSDEVIC in unquoted form. To quote the resolved value of SYSDEVIC, use the %SUPERQ macro quoting function.

Comparisons

Assigning a value to SYSDEVIC is the same as specifying a value for the DEVICE= system option.

SYSDMG Automatic Macro Variable

Contains a return code that reflects an action taken on a damaged data set.

Type: Automatic macro variable (read and write)

Default: 0

Details

You can use the value of SYSDMG as a condition to determine further action to take.

SYSDMG can contain the following values:

Table 14.2 *SYSDMG Values and Descriptions*

Value	Description
0	No repair of damaged data sets in this session. (Default)
1	One or more automatic repairs of damaged data sets has occurred.
2	One or more user-requested repairs of damaged data sets has occurred.
3	One or more opens failed because the file was damaged.
4	One or more SAS tasks were terminated because of a damaged data set.

Value	Description
5	One or more automatic repairs of damaged data sets has occurred; the last-repaired data set has index file removed, as requested.
6	One or more user requested repairs has occurred; the last-repaired data set has index file removed, as requested.

SYSDSN Automatic Macro Variable

Contains the libref and name of the most recently created SAS data set.

Type: Automatic macro variable (read and write)

See: "SYSLAST Automatic Macro Variable" on page 204

Details

The libref and data set name are displayed in two left-aligned fields. If no SAS data set has been created in the current program, SYSDSN returns eight blanks followed by _NULL_ followed by two more blanks.

Note: The macro processor always stores the value of SYSDSN in unquoted form. To quote the resolved value of SYSDSN, use the %SUPERQ macro quoting function.

Comparisons

- Assigning a value to SYSDSN is the same as specifying a value for the _LAST_= system option.
- The value of SYSLAST is often more useful than SYSDSN because the value of SYSLAST is formatted so that you can insert a reference to it directly into SAS code in place of a data set name.

Example: Comparing Values Produced by SYSDSN and SYSLAST

Create a data set WORK.TEST and then enter the following statements:

```
%put Sysdsn produces:  *&sysdsn*;
%put Syslast produces: *&syslast*;
```

When these statements execute, the following lines are written to the SAS log:

```
Sysdsn produces:  *WORK    TEST    *
Syslast produces: *WORK.TEST     *
```

When the libref or data set name contain fewer than eight characters, SYSDSN maintains the blanks for the unused characters. SYSDSN does not display a period between the libref and data set name fields.

SYSENCODING Automatic Macro Variable

Contains the name of the SAS session encoding.

Type: Automatic macro variable (ready only)

Details

SYSENCODING displays the name with a maximum length of 12 bytes.

Example: Using SYSENCODING to Display the SAS Session Encoding

The following statement displays the encoding for the SAS session:

```
%put The encoding for this SAS session is: &sysencoding;
```

When this statement executes, the following comment is written to the SAS log:

```
The encoding for this SAS session is: wlatin1
```

SYSENDIAN Automatic Macro Variable

Contains an indication of the byte order of the current session. The possible values are LITTLE or BIG.

Type: Automatic macro variable (read only)

Details

The SYSENDIAN automatic macro variable indicates the byte order of the current SAS session. There are two possible values: LITTLE and BIG.

SYSENV Automatic Macro Variable

Reports whether SAS is running interactively.

Type: Automatic macro variable (read only)

Details

The value of SYSENV is independent of the source of input. The following are values for SYSENV:

FORE
: when the SAS system option TERMINAL is in effect. For example, the value is FORE when you run SAS interactively through a windowing environment.

BACK
: when the SAS system option NOTERMINAL is in effect. For example, the value is BACK when you submit a SAS job in batch mode.

You can use SYSENV to check the execution mode before submitting code that requires interactive processing. To use a %INPUT statement, the value of SYSENV must be FORE. For details, see the SAS documentation for your operating environment.

Operating Environment Information
Some operating environments do not support the submission of jobs in batch mode. In this case the value of SYSENV is always FORE. For details, see the SAS documentation for your operating environment.

SYSERR Automatic Macro Variable

Contains a return code status set by some SAS procedures and the DATA step.

Type: Automatic macro variable (read only)

Details

You can use the value of SYSERR as a condition to determine further action to take or to decide which parts of a SAS program to execute. SYSERR is used to detect major system errors, such as out of memory or failure of the component system when used in some procedures and DATA steps. SYSERR automatic macro variable is reset at each step boundary. For the return code of a complete job, see “SYSCC Automatic Macro Variable” on page 193.

SYSERR can contain the following values:

Table 14.3 *SYSERR Values*

Value	Description
0	Execution completed successfully and without warning messages.
1	Execution was canceled by a user with a RUN CANCEL statement.
2	Execution was canceled by a user with an ATTN or BREAK command.
3	An error in a program run in batch or non-interactive mode caused SAS to enter syntax-check mode.
4	Execution completed successfully but with warning messages.
5	Execution was canceled by a user with an ABORT CANCEL statement.
6	Execution was canceled by a user with an ABORT CANCEL FILE statement.
>6	An error occurred. The value returned is procedure-dependent.

The following table contains warning return codes. The codes do not indicate any specific problems. These codes are guidelines to identify the nature of a problem.

Table 14.4 SYSERR Warning Codes

Warning Code	Description
108	Problem with one or more BY groups
112	Error with one or more BY groups
116	Memory problems with one or more BY groups
120	I/O problems with one or more BY groups

The following table contains error return codes. The codes do not indicate any specific problems. These codes are guidelines to identify the nature of a problem.

Table 14.5 SYSERR Error Codes

Error Code	Description
1008	General data problem
1012	General error condition
1016	Out-of-memory condition
1020	I/O problem
2000	Semantic action problem
2001	Attribute processing problem
3000	Syntax error
4000	Not a valid procedure
9999	Bug in the procedure
20000	A step was stopped or an ABORT statement was issued.
20001	An ABORT RETURN statement was issued.
20002	An ABORT ABEND statement was issued.
25000	Severe system error. The system cannot initialize or continue.

Example: Using SYSERR

The example creates an error message and uses %PUT &SYSERR to write the return code number (1012) to the SAS log.

```
data NULL;
   set doesnotexist;
```

```
run;
%put &syserr;
```

The following SAS log output contains the return code number:

```
75   data NULL;
76      set doesnotexist;
ERROR: File WORK.DOESNOTEXIST.DATA does not exist.
77
78   run;
NOTE: The SAS System stopped processing this step because of errors.
WARNING: The data set WORK.NULL might be incomplete.  When this step was
stopped there were 0 observations and 0 variables.
WARNING: Data set WORK.NULL was not replaced because this step was stopped.
NOTE: DATA statement used (Total process time):
      real time           0.00 seconds
      cpu time            0.00 seconds
79
80   %put &syserr;
1012
```

To retrieve error and warning text instead of the return code number, see “SYSERRORTEXT Automatic Macro Variable” on page 202 and “SYSWARNINGTEXT Automatic Macro Variable” on page 219.

SYSERRORTEXT Automatic Macro Variable

Contains the text of the last error message formatted for display in the SAS log.

Type: Automatic macro variable (read only)

Details

The value of SYSERRORTEXT is the text of the last error message generated in the SAS log. For a list of SYSERR warnings and errors, see “SYSERR Automatic Macro Variable” on page 200.

Note: If the last error message text that was generated contains an `&` or `%` and you are using the %PUT statement, you must use the %SUPERQ macro quoting function to mask the special characters to prevent further resolution of the value. The following example uses the %PUT statement and the %SUPERQ macro quoting function:

```
%put %superq(syserrortext);
```

For more information, see “%SUPERQ Function” on page 259.

Example: Using SYSERRORTEXT

This example creates an error message:

```
data NULL;
   set doesnotexist;
run;
%put &syserrortext;
```

When these statements are executed, the following record is written to the SAS log:

```
1   data NULL;
2   set doesnotexist;
ERROR: File WORK.DOESNOTEXIST.DATA does not exist.
3   run;
NOTE: The SAS System stopped processing this step because of errors.
WARNING: The data set WORK.NULL might be incomplete. When this step was
         stopped there were 0 observations and 0 variables.
NOTE: DATA statement used (Total process time):
real time        11.16 seconds
cpu time         0.07 seconds
4  %put &syserrortext;
File WORK.DOESNOTEXIST.DATA does not exist.
```

SYSFILRC Automatic Macro Variable

Contains the return code from the last FILENAME statement.

Type: Automatic macro variable (read and write)

Details

SYSFILRC checks whether the file or storage location referenced by the last FILENAME statement exists. You can use SYSFILRC to confirm that a file or location is allocated before attempting to access an external file.

The following are values for SYSFILRC:

Table 14.6 *SYSFILRC Values and Descriptions*

Value	Description
0	The last FILENAME statement executed correctly.
≠0	The last FILENAME statement did not execute correctly.

SYSHOSTNAME Automatic Macro Variable

Contains the host name of a computer.

Type: Automatic macro variable (read only)

Details

SYSHOSTNAME contains the host name of the system that is running a single TCPIP stack. For more information about TCPIP stacks, see your SAS host companion documentation.

SYSINDEX Automatic Macro Variable

Contains the number of macros that have started execution in the current SAS job or session.

Type: Automatic macro variable (read only)

Details

You can use SYSINDEX in a program that uses macros when you need a unique number that changes after each macro invocation.

SYSINFO Automatic Macro Variable

Contains return codes provided by some SAS procedures.

Type: Automatic macro variable (read only)

Details

Values of SYSINFO are described with the procedures that use it. You can use the value of SYSINFO as a condition for determining further action to take or parts of a SAS program to execute.

For example, PROC COMPARE, which compares two data sets, uses SYSINFO to store a value that provides information about the result of the comparison.

SYSJOBID Automatic Macro Variable

Contains the name of the current batch job or user ID.

Type: Automatic macro variable (read only)

Details

The value stored in SYSJOBID depends on the operating environment that you use to run SAS. You can use SYSJOBID to check who is currently executing the job to restrict certain processing or to issue commands that are specific to a user.

SYSLAST Automatic Macro Variable

Contains the name of the SAS data file created most recently.

Type: Automatic macro variable (read and write)

See: "SYSDSN Automatic Macro Variable" on page 198

Details

The name is stored in the form *libref.dataset*. You can insert a reference to SYSLAST directly into SAS code in place of a data set name. If no SAS data set has been created in the current program, the value of SYSLAST is NULL , with no leading or trailing blanks.

Note: The macro processor always stores the value of SYSLAST in unquoted form. To quote the resolved value of SYSLAST, use the %SUPERQ macro quoting function.

Comparisons

- Assigning a value to SYSLAST is the same as specifying a value for the _LAST_= system option.
- The value of SYSLAST is often more useful than SYSDSN because the value of SYSLAST is formatted so that you can insert a reference to it directly into SAS code in place of a data set name.

Example: Comparing Values Produced by SYSLAST and SYSDSN

Create the data set FIRSTLIB.SALESRPT and then enter the following statements:

```
%put Sysdsn produces:  *&sysdsn*;
%put Syslast produces: *&syslast*;
```

When these statements are executed, the following is written to the SAS log:

```
Sysdsn produces:  *FIRSTLIBSALESRPT*
Syslast produces: *FIRSTLIB.SALESRPT*
```

The name stored in SYSLAST contains the period between the libref and data set name.

SYSLCKRC Automatic Macro Variable

Contains the return code from the most recent LOCK statement.

Type: Automatic macro variable (read and write)

Details

The LOCK statement is a Base SAS software statement used to acquire and release an exclusive lock on data objects in data libraries accessed through SAS/SHARE software. The following are values for SYSLCKRC:

Table 14.7 *LCKRC Values and Descriptions*

Value	Description
0	The last LOCK statement was successful.
>0	The last LOCK statement was not successful.

Value	Description
<0	The last LOCK statement was completed, but a WARNING or NOTE was written to the SAS log.

For more information, see the documentation for SAS/SHARE software.

SYSLIBRC Automatic Macro Variable

Contains the return code from the last LIBNAME statement.

Type: Automatic macro variable (read and write)

Details

The code reports whether the last LIBNAME statement executed correctly. SYSLIBRC checks whether the SAS library referenced by the last LIBNAME statement exists. For example, you could use SYSLIBRC to confirm that a libref is allocated before you attempt to access a permanent data set.

The following are values for SYSLIBRC:

Table 14.8 *SYSLIBRC Values and Descriptions*

Value	Description
0	The last LIBNAME statement executed correctly.
≠0	The last LIBNAME statement did not execute correctly.

SYSLOGAPPLNAME Automatic Macro Variable

Contains the value of the LOGAPPLNAME= system option.

Type: Automatic macro variable (read only)

Default: null

Details

The following code, when submitted from the current SAS session, writes the LOGAPPLNAME for the current SAS session to the log:

```
%put &syslogapplname;
```

SYSMACRONAME Automatic Macro Variable

Returns the name of the currently executing macro.

Type: Automatic macro variable (read only)

Details

When referenced outside of an executing macro, SYSMACRONAME returns the null string.

SYSMENV Automatic Macro Variable

Contains the invocation status of the macro that is currently executing.

Type: Automatic macro variable (read only)

Details

The following are values for SYSMENV:

Table 14.9 *SMENV Values and Descriptions*

Value	Description
S	The macro currently executing was invoked as part of a SAS program.
D	The macro currently executing was invoked from the command line of a SAS window.

SYSMSG Automatic Macro Variable

Contains text to display in the message area of a macro window.

Type: Automatic macro variable (read and write)

Details

Values assigned to SYSMSG do not require quotation marks. The value of SYSMSG is set to null after each execution of a %DISPLAY statement.

Example: %DISPLAY Statement

This example shows that text assigned to SYSMSG is cleared after the %DISPLAY statement.

```
%let sysmsg=Press ENTER to continue.;
%window start
   #5  @28 'Welcome to SAS';
%display start;
%put Sysmsg is: *&sysmsg*;
```

When this program executes, the following is written to the SAS log:

```
Sysmsg is: **
```

SYSNCPU Automatic Macro Variable

Contains the current number of processors available to SAS for computations.

Type: Automatic Macro Variable (Read Only)

Details

SYSNCPU is an automatic macro variable that provides the current value of the CPUCOUNT option. For more information about "CPUCOUNT= System Option" in *SAS System Options: Reference.*

Comparisons

The following example shows the option CPUCOUNT set to 265.

```
options cpucount=265;
%put &sysncpu;
```

The output of the above example is 265.

SYSNOBS Automatic Macro Variable

Contains the number of observations read from the last data set that was closed by the previous procedure or DATA step.

Type: Automatic macro variable (read only)

Details

The SYSNOBS contains the number of observations read from the last data set that was closed by the previous procedure or DATA step.

Note: If the number of observations for the data set was not calculated by the previous procedure or DATA step, the value of SYSNOBS is set to -1.

SYSODSESCAPECHAR Automatic Macro Variable

Displays the value of the ODS ESCAPECHAR= from within the program.

Type: Automatic macro variable (read only)

Details

SYSODSESCAPECHAR automatic macro variable contains the current ODS escape character.

SYSODSPATH Automatic Macro Variable

Contains the current Output Delivery System (ODS) pathname.

Type: Automatic macro variable (read only)

Restriction: The SYSODSPATH automatic macro variable exists only when an ODS or PROC TEMPLATE statement is invoked.

Details

The SYSODSPATH automatic macro variable contains the current ODS pathname.

SYSPARM Automatic Macro Variable

Contains a character string that can be passed from the operating environment to SAS program steps.

Type: Automatic macro variable (read and write)

Details

SYSPARM enables you to pass a character string from the operating environment to SAS program steps and provides a means of accessing or using the string while a program is executing. For example, you can use SYSPARM from the operating environment to pass a title statement or a value for a program to process. You can also set the value of SYSPARM within a SAS program. SYSPARM can be used anywhere in a SAS program. The default value of SYSPARM is null (zero characters).

SYSPARM is most useful when specified at invocation of SAS. For details, see the SAS documentation for your operating environment.

Note: The macro processor always stores the value of SYSPARM in unquoted form. To quote the resolved value of SYSPARM, use the %SUPERQ macro quoting function.

Comparisons

- Assigning a value to SYSPARM is the same as specifying a value for the SYSPARM= system option.
- Retrieving the value of SYSPARM is the same as using the SYSPARM() SAS function.

Example: Passing a Value to a Procedure

In this example, you invoke SAS on a UNIX operating environment on September 20, 2011 (the librefs DEPT and TEST are defined in the config.sas file) with a command like the following:

```
sas program-name -sysparm dept.projects -config /myid/config.sas
```

Macro variable SYSPARM supplies the name of the data set for PROC REPORT:

```
proc report data=&sysparm
     report=test.resorces.priority.rept;
```

```
title "%sysfunc(date(),worddate.)";
title2;
title3 'Active Projects By Priority';
run;
```

SAS sees the following:

```
proc report data=dept.projects
     report=test.resorces.priority.rept;
title "September 20, 2011";
title2;
title3 'Active Projects By Priority';
run;
```

SYSPBUFF Automatic Macro Variable

Contains text supplied as macro parameter values.

Type: Automatic macro variable (read and write, local scope)

Details

SYSPBUFF resolves to the text supplied as parameter values in the invocation of a macro that is defined with the PARMBUFF option. For name-style invocations, this text includes the parentheses and commas. Using the PARMBUFF option and SYSPBUFF, you can define a macro that accepts a varying number of parameters at each invocation.

If the macro definition includes both a set of parameters and the PARMBUFF option, the macro invocation causes the parameters to receive values and the entire invocation list of values to be assigned to SYSPBUFF.

Example: Using SYSPBUFF to Display Macro Parameter Values

The macro PRINTZ uses the PARMBUFF option to define a varying number of parameters and SYSPBUFF to display the parameters specified at invocation.

```
%macro printz/parmbuff;
   %put Syspbuff contains: &syspbuff;
   %let num=1;
   %let dsname=%scan(&syspbuff,&num);
   %do %while(&dsname ne);
      proc print data=&dsname;
      run;
      %let num=%eval(&num+1);
      %let dsname=%scan(&syspbuff,&num);
   %end;
%mend printz;
%printz(purple,red,blue,teal)
```

When this program executes, this line is written to the SAS log:

```
Syspbuff contains: (purple,red,blue,teal)
```

SYSPROCESSID Automatic Macro Variable

Contains the process ID of the current SAS process.

Type: Automatic macro variable (read only)

Default: null

Details

The process ID is a 32–character hexadecimal string. The default value is null.

Example: Using SYSPROCESSID to Display the Current SAS Process ID

The following code writes the current SAS process ID to the SAS log:

```
%put &sysprocessid;
```

A process ID, such as the following, is written to the SAS log:

```
41D1B269F86C7C5F4010000000000000
```

SYSPROCESSNAME Automatic Macro Variable

Contains the process name of the current SAS process.

Type: Automatic macro variable (read only)

Example: Using SYSPROCESSNAME to Display the Current SAS Process Name

The following statement writes the name of the current SAS process to the log:

```
%put &sysprocessname;
```

If you submit this statement in the SAS windowing environment of your second SAS session, the following line is written to the SAS log:

```
DMS Process (2)
```

SYSPROCNAME Automatic Macro Variable

Contains the name of the procedure (or DATASTEP for DATA steps) currently being processed by the SAS Language Processor.

Type: Automatic macro variable (read only)

Details

The value of SYSPROCNAME contains the name of the procedure specified by the user in the PROC statement until a step boundary is reached.

SYSRC Automatic Macro Variable

Contains the last return code generated by your operating system.

Type: Automatic macro variable (read and write)

Details

The code returned by SYSRC is based on commands that you execute using the X statement in open code, the X command in a windowing environment, or the %SYSEXEC, %TSO, or %CMS macro statements. Return codes are integers. The default value of SYSRC is 0.

You can use SYSRC to check the return code of a system command before you continue with a job. For return code examples, see the SAS companion for your operating environment.

SYSSCP and SYSSCPL Automatic Macro Variables

Contain an identifier for your operating environment.

Type: Automatic macro variable (read only)

Details

SYSSCP and SYSSCPL resolve to an abbreviation of the name of your operating environment. In some cases, SYSSCPL provides a more specific value than SYSSCP. You could use SYSSCP and SYSSCPL to check the operating environment to execute appropriate system commands.

The following table lists the values for SYSSCP and SYSSCPL.

Table 14.10 *SYSSCP and SYSSCPL Values for Platforms Running SAS 9.2 or Later*

Platform	SYSSCP Value	SYSSCPL Value
z/OS	OS	z/OS
VMI or OpenVMS on Itanium - supported for Foundation SAS only	VMS ITAN	OpenVMS
	UNIX	
HP-UX PA-RISC or H64	HP 64	HP-UX
H61 or HP-UX IPF or HP-UX on Itanium	HP IPF	HP-UX

Platform	SYSSCP Value	SYSSCPL Value
LAX or LINUX on X64 (x86-64)	LIN X64	LINUX
LNX or LINUX or LINUX 32-bit (x86)	LINUX	LINUX
R64 or AIX64 or AIX on POWER	AIX 64	AIX
S64 or SUN64 or Solaris on SPARC	SUN 64	SUNOS or SunOS
SAX or Solaris 10 on X64 (x86-64)	SUN X64	SUNOS
	Windows	
Windows XP Pro	WIN	XP_PRO
Windows Server 2003	WIN	NET_SRV
Windows Enterprise Server 2003	WIN	NET_ASRV
Windows Data Center Server 2003	WIN	NET_DSRV
Windows XP Pro x64	WIN	X64_PRO
Windows Server 2003 x64	WIN	X64_SRV
Windows Enterprise Server 2003 x64	WIN	X64_ESRV
Windows Data Center Server 2003 x64	WIN	X64_DSRV
Windows Vista Business	WIN	W32_VSPRO
Windows Server 2008	WIN	W32_SRV08
Windows Enterprise Server 2008	WIN	W32_ESRV08
Windows Data Center Server 2008	WIN	W32_DSRV08
Windows Vista Business x64	WIN	X64_VSPRO
Windows Server 2008 x64	WIN	X64_SRV08
Windows Enterprise Server 2008 x64	WIN	X64_ESRV08

Platform	SYSSCP Value	SYSSCPL Value
Windows Data Center Server 2008 x64	WIN	X64_DSRV08
Windows Server 2008 Itanium	WIN	W64_ESRV08
Windows Itanium Enterprise Server 2003 or W64_ASRV	WIN	W64_ASRV
Windows Itanium Data Center Server 2003 or W64_DSRV	WIN	W64_DSRV
Windows Itanium Server 2003	WIN	W64_SRV

Example: Deleting a Temporary File on a Platform Running SAS

The macro DELFILE locates the platform that is running SAS and deletes the TMP file. FILEREF is a global macro variable that contains the fileref for the TMP file.

```
%macro delfile;
   %if  /* HP Unix */&sysscp=HP 800 or &sysscp=HP 300
   %then
      %do;
         X "rm &fileref..TMP";
      %end;
      %else %if  /* DOS-LIKE PLATFORMS */&sysscp=OS2 or &sysscp=WIN
   %then
      %do;
          X "DEL &fileref..TMP";
      %end;
   %else %if  /* CMS */&sysscp=CMS
   %then
      %do;
         X "ERASE &fileref TEMP A";
      %end;
%mend delfile;
```

SYSSCPL Automatic Macro Variable

Contains the name of your operating environment.

Type: Automatic macro variable (read only)

Details

See “SYSSCP and SYSSCPL Automatic Macro Variables” on page 212.

SYSSITE Automatic Macro Variable

Contains the number assigned to your site.

Type: Automatic macro variable (read only)

Details

SAS assigns a site number to each site that licenses SAS software. The number displays in the SAS log.

SYSSIZEOFLONG Automatic Macro Variable

Contains the length in bytes of a long integer in the current session.

Type: Automatic macro variable (read only)

Details

The SYSSIZEOFLONG automatic macro variable contains the length of a long integer in the current SAS session.

SYSSIZEOFPTR Automatic Macro Variable

Contains the size in bytes of a pointer.

Type: Automatic macro variable (read only)

Details

The SYSSIZEOFPTR automatic macro variable contains the size in bytes of a pointer.

SYSSIZEOFUNICODE Automatic Macro Variable

Contains the length in bytes of a Unicode character in the current session.

Type: Automatic macro variable (read only)

Details

The SYSSIZEOFUNICODE automatic macro variable contains the length of the Unicode character in the current SAS session.

SYSSTARTID Automatic Macro Variable

Contains the ID generated from the last STARTSAS statement (experimental).

Type: Automatic macro variable (read only)

Default: null

Note: STARTSAS statement is an experimental feature of the SAS System.

Details

The ID is a 32-character hexadecimal string that can be passed to the WAITSAS statement or the ENDSAS statement. The default value is null.

Example: Using SYSSTARTID to Display the SAS Process ID from the Most Recent STARTSAS Statement (experimental)

Submit the following code from the SAS process in which you have submitted the most recent STARTSAS statement to write the value of the SYSSTARTID variable to the SAS log:

```
%put &sysstartid
```

A process ID value, such as the following, is written to the SAS log:

```
41D20425B89FCED94036000000000000
```

SYSSTARTNAME Automatic Macro Variable

Contains the process name generated from the last STARTSAS statement (experimental).

Type: Automatic macro variable (read only)

Default: null

Note: STARTSAS statement is an experimental feature of the SAS System.

Example: Using SYSSTARTNAME to Display the SAS Process Name from the Most Recent STARTSAS Statement (experimental)

Submit the following code from the SAS process in which you have submitted the most recent STARTSAS statement to write the value of the SYSSTARTNAME variable to the SAS log:

```
%put &sysstartname;
```

An example of a process name that can appear in the SAS log is as follows:

```
DMS Process (2)
```

SYSTCPIPHOSTNAME Automatic Macro Variable

Contains the host names of the local and remote computers when multiple TCP/IP stacks are supported.

Type: Automatic macro variable (read only)

Details

SYSTCPIPHOSTNAME contains the host name of the system that is running multiple TCPIP stacks. For more information about TCPIP stacks, see your SAS host companion documentation.

SYSTIME Automatic Macro Variable

Contains the time a SAS job or session began executing.

Type: Automatic macro variable (read only)

Details

The value is displayed in TIME5. format and does not change during the individual job or session.

Example: Using SYSTIME to Display the Time that a SAS Session Started

The following statement displays the time a SAS session started.

```
%put This SAS session started running at: &systime;
```

When this statement executes at 3 p.m., but your SAS session began executing at 9:30 a.m., the following comment is written to the SAS log:

```
This SAS session started running at: 09:30
```

SYSUSERID Automatic Macro Variable

Contains the user ID or login of the current SAS process.

Type: Automatic macro variable (read only)

Example: Using SYSUSERID to Display the User ID for the Current SAS Process

The following code, when submitted from the current SAS process, writes the user ID or login for the current SAS process to the SAS log:

```
%put &sysuserid;
```

A user ID, such as the following, is written to the SAS log:

```
MyUserid
```

SYSVER Automatic Macro Variable

Contains the release number of SAS software that is running.

Type: Automatic macro variable (read only)

See: "SYSVLONG Automatic Macro Variable" on page 218 and "SYSVLONG4 Automatic Macro Variable" on page 218

Comparisons

SYSVER provides the release number of the SAS software that is running. You can use SYSVER to check for the release of SAS before running a job with newer features.

Example: Identifying SAS Software Release

The following statement displays the release number of a user's SAS software.

```
%put I am using release: &sysver;
```

Submitting this statement (for a user of SAS 9.2) writes the following to the SAS log:

```
I am using release: 9.2
```

SYSVLONG Automatic Macro Variable

Contains the release number and maintenance level of SAS software that is running.

Type: Automatic macro variable (read only)

See: "SYSVER Automatic Macro Variable" on page 218 and "SYSVLONG4 Automatic Macro Variable" on page 218

Comparisons

SYSVLONG provides the release number and maintenance level of SAS software, in addition to the release number.

Example: Identifying a SAS Maintenance Release

The following statement displays information identifying the SAS release being used.

```
%put I am using release: &sysvlong;
```

Submitting this statement (for a user of SAS 9.2) writes the following to the SAS log:

```
I am using release: 9.02.02M2D071609
```

SYSVLONG4 Automatic Macro Variable

Contains the release number and maintenance level of SAS software that is running and has a four digit year.

Type: Automatic macro variable (read only)

See: "SYSVER Automatic Macro Variable" on page 218 and "SYSVLONG Automatic Macro Variable" on page 218

Comparisons

SYSVLONG4 provides a four digit year and the release number and maintenance level of SAS software. SYSVLONG does not contain the four digit year but everything else is the same.

Example: Using SYSVLONG4 Automatic Macro Variable

The following statement displays information that identifies the SAS release being used.

```
%put I am using maintenance release: &sysvlong4;
```

Submitting this statement (for a user of SAS 9.2) writes this comment to the SAS log:

```
I am using maintenance release: 9.02.01B0D09112007
```

SYSWARNINGTEXT Automatic Macro Variable

Contains the text of the last warning message formatted for display in the SAS log.

Type: Automatic macro variable (read only)

Details

The value of SYSWARNINGTEXT is the text of the last warning message generated in the SAS log. For a list of SYSERR warnings and errors, see "SYSERR Automatic Macro Variable" on page 200.

Note: If the last warning message text that was generated contains an & or % and you are using the %PUT statement, you must use the %SUPERQ macro quoting function to mask the special characters to prevent further resolution of the value. The following example uses the %PUT statement and the %SUPERQ macro quoting function:

```
%put %superq(syswarningtext);
```

For more information, see "%SUPERQ Function" on page 259.

Example: Using SYSWARNINGTEXT

This example creates a warning message:

```
data NULL;
    set doesnotexist;
run;
%put &syswarningtext;
```

When these statements execute, the following comments are written to the SAS log:

```
1  data NULL;
2  set doesnotexist;
ERROR: File WORK.DOESNOTEXIST.DATA does not exist.
3  run;
NOTE: The SAS System stopped processing this step because of errors.
WARNING: The data set WORK.NULL might be incomplete.  When this step
         was stopped there were 0 observations and 0 variables.
NOTE: DATA statement used (Total process time):
    real time           11.16 seconds
    cpu time            0.07 seconds
4  %put &syswarningtext;
The data set WORK.NULL might be incomplete.  When this step was
stopped there were 0 observations and 0 variables.
```

Chapter 15
DATA Step Call Routines for Macros

DATA Step Call Routines for Macros

You can interact with the Macro Facility using DATA step call routines.

Dictionary

CALL EXECUTE Routine

Resolves the argument, and issues the resolved value for execution at the next step boundary.

Type: DATA step call routine

Syntax

CALL EXECUTE (*argument*);

Required Argument

argument
can be one of the following:

- a character string, enclosed in quotation marks. *Argument* within single quotation marks resolves during program execution. *Argument* within double quotation marks resolves while the DATA step is being constructed. For example, to invoke the macro SALES, you can use the following code:

  ```
  call execute('%sales');
  ```

- the name of a DATA step character variable whose value is a text expression or a SAS statement to be generated. Do not enclose the name of the DATA step variable in quotation marks. For example, to use the value of the DATA step variable FINDOBS, which contains a SAS statement or text expression, you can use the following code:

```
call execute(findobs);
```

- a character expression that is resolved by the DATA step to a macro text expression or a SAS statement. For example, to generate a macro invocation whose parameter is the value of the variable MONTH, you use the following code:

```
call execute('%sales('||month||')');
```

Details

If an EXECUTE routine argument is a macro invocation or resolves to one, the macro executes immediately. Execution of SAS statements generated by the execution of the macro will be delayed until after a step boundary. SAS macro statements, including macro variable references, will execute immediately.

Note: Because of the delay of the execution of the SAS statements until after a step boundary, references in SAS macro statements to macro variables created or updated by the SAS statements will not resolve properly.

Note: Because macro references execute immediately and SAS statements do not execute until after a step boundary, you cannot use CALL EXECUTE to invoke a macro that contains references for macro variables that are created by CALL SYMPUT in that macro. For a workaround, see the following TIP.

TIP The following example uses the %NRSTR macro quoting function to mask the macro statement. This function will delay the execution of macro statements until after a step boundary.

```
call execute('%nrstr(%sales('||month||'))');
```

Comparisons

Unlike other elements of the macro facility, a CALL EXECUTE statement is available regardless of the setting of the SAS system option MACRO|NOMACRO. In both cases, EXECUTE places the value of its argument in the program stack. However, when NOMACRO is set, any macro calls or macro functions in the argument are not resolved.

Examples

Example 1: Executing a Macro Conditionally

The following DATA step uses CALL EXECUTE to execute a macro only if the DATA step writes at least one observation to the temporary data set.

```
%macro overdue;
   proc print data=late;
      title "Overdue Accounts As of &sysdate";
   run;
%mend overdue;
data late;
   set sasuser.billed end=final;
   if datedue<=today()-30 then
```

```
      do;
         n+1;
         output;
      end;
   if final and n then call execute('%overdue');
run;
```

Example 2: Passing DATA Step Values into a Parameter List

CALL EXECUTE passes the value of the DATE variable in the DATES data set to macro REPT for its DAT parameter, the value of the VAR1 variable in the REPTDATA data set for its A parameter, and REPTDATA as the value of its DSN parameter. After the DATA _NULL_ step finishes, three PROC GCHART statements are submitted, one for each of the three dates in the DATES data set.

```
data dates;
   input date $;
datalines;
10nov11
11nov11
12nov11
;
data reptdata;
   input date $ var1 var2;
datalines;
10nov11 25 10
10nov11 50 11
11nov11 23 10
11nov11 30 29
12nov11 33 44
12nov11 75 86
;
%macro rept(dat,a,dsn);
   proc chart data=&dsn;
       title "Chart for &dat";
       where(date="&dat");
       vbar &a;
   run;
%mend rept;
data _null_;
   set dates;
   call execute('%rept('||date||','||'var1,reptdata)');
run;
```

CALL SYMDEL Routine

Deletes the specified variable from the macro global symbol table.

Type: DATA step call routine

Syntax

CALL SYMDEL(*macro-variable*<, *option*>);

Required Arguments

macro-variable
can be any of the following:

- the name of a macro variable within quotation marks but without an ampersand. When a macro variable value contains another macro variable reference, SYMDEL does not attempt to resolve the reference.
- the name of a DATA step character variable, specified with no quotation marks, which contains the name of a macro variable. If the value is not a valid SAS name, or if the macro processor cannot find a macro variable of that name, SAS writes a warning to the log.
- a character expression that constructs a macro variable name.

option(s)

NOWARN
suppresses the warning message when an attempt is made to delete a non-existent macro variable. NOWARN must be within quotation marks.

Details

CALL SYMDEL issues a warning when an attempt is made to delete a non-existent macro variable. To suppress this message, use the NOWARN option.

CALL SYMPUT Routine

Assigns a value produced in a DATA step to a macro variable.

Type: DATA step call routine

See: "SYMGET Function" on page 234 and "CALL SYMPUTX Routine" on page 229

Syntax

CALL SYMPUT(*macro-variable*, *value*);

Required Arguments

macro-variable
can be one of the following items:

- a character string that is a SAS name, enclosed in quotation marks. For example, to assign the character string **testing** to macro variable NEW, submit the following statement:

```
call symput('new','testing');
```

- the name of a character variable whose values are SAS names. For example, this DATA step creates the three macro variables SHORTSTP, PITCHER, and FRSTBASE and respectively assign them the values ANN, TOM, and BILL.

```
data team1;
   input position : $8. player : $12.;
   call symput(position,player);
datalines;
shortstp Ann
```

```
pitcher Tom
frstbase Bill
;
```

- a character expression that produces a macro variable name. This form is useful for creating a series of macro variables. For example, the CALL SYMPUT statement builds a series of macro variable names by combining the character string POS and the left-aligned value of _N_ and assigns values to the macro variables POS1, POS2, and POS3.

```
data team2;
   input position : $12. player $12.;
   call symput('POS'||left(_n_), position);
   datalines;
shortstp Ann
pitcher Tom
frstbase Bill
;
```

value

is the value to be assigned, which can be

- a string enclosed in quotation marks. For example, this statement assigns the string **testing** to the macro variable NEW:

```
call symput('new','testing');
```

- the name of a numeric or character variable. The current value of the variable is assigned as the value of the macro variable. If the variable is numeric, SAS performs an automatic numeric-to-character conversion and writes a message in the log. Later sections on formatting rules describe the rules that SYMPUT follows in assigning character and numeric values of DATA step variables to macro variables.

  Note: This form is most useful when *macro-variable* is also the name of a SAS variable or a character expression that contains a SAS variable. A unique macro variable name and value can be created from each observation, as shown in the previous example for creating the data set TEAM1.

  If *macro-variable* is a character string, SYMPUT creates only one macro variable, and its value changes in each iteration of the program. Only the value assigned in the last iteration remains after program execution is finished.

- a DATA step expression. The value returned by the expression in the current observation is assigned as the value of *macro-variable.* In this example, the macro variable named HOLDATE receives the value **July 4,1997**:

```
data c;
   input holiday mmddyy.;
   call symput('holdate',trim(left(put(holiday,worddate.))));
datalines;
070497
;
run;
```

  If the expression is numeric, SAS performs an automatic numeric-to-character conversion and writes a message in the log. Later sections on formatting rules describe the rules that SYMPUT follows in assigning character and numeric values of expressions to macro variables.

Details

If *macro-variable* does not exist, SYMPUT creates it. SYMPUT makes a macro variable assignment when the program executes.

SYMPUT can be used in all SAS language programs, including SCL programs. Because it resolves variables at program execution instead of macro execution, SYMPUT should be used to assign macro values from DATA step views, SQL views, and SCL programs.

Scope of Variables Created with SYMPUT

SYMPUT puts the macro variable in the most local nonempty symbol table. A symbol table is nonempty if it contains the following:

- a value
- a computed %GOTO (A computed %GOTO contains `%`or `&` and resolves to a label.)
- the macro variable &SYSPBUFF, created at macro invocation time.

However, there are three cases where SYMPUT creates the variable in the local symbol table, even if that symbol table is empty:

- Beginning with Version 8, if SYMPUT is used after a PROC SQL, the variable will be created in a local symbol table.
- If an executing macro contains a computed %GOTO statement and uses SYMPUT to create a macro variable, the variable is created in the local symbol table.
- If an executing macro uses &SYSPBUFF and SYMPUT to create a macro variable, the macro variable is created in the local symbol table.

For more information about creating a variable with SYMPUT, see "Scopes of Macro Variables" on page 43.

Problem Trying to Reference a SYMPUT-Assigned Value Before It Is Available

One of the most common problems in using SYMPUT is trying to reference a macro variable value assigned by SYMPUT before that variable is created. The failure generally occurs because the statement referencing the macro variable compiles before execution of the CALL SYMPUT statement that assigns the variable's value. The most important fact to remember in using SYMPUT is that it assigns the value of the macro variable during program execution. Macro variable references resolve during the compilation of a step, a global statement used outside a step, or an SCL program. As a result:

- You cannot use a macro variable reference to retrieve the value of a macro variable in the same program (or step) in which SYMPUT creates that macro variable and assigns it a value.
- You must specify a step boundary statement to force the DATA step to execute before referencing a value in a global statement following the program (for example, a TITLE statement). The boundary could be a RUN statement or another DATA or PROC statement. For example:

```
data x;
   x='December';
   call symput('var',x);
proc print;
title "Report for &var";
run;
```

Processing on page 33 provides details about compilation and execution.

Formatting Rules For Assigning Character Values

If *value* is a character variable, SYMPUT writes it using the $*w*. format, where *w* is the length of the variable. Therefore, a value shorter than the length of the program variable is written with trailing blanks. For example, in the following DATA step the length of variable C is 8 by default. Therefore, SYMPUT uses the $8. format and assigns the letter **x** followed by seven trailing blanks as the value of CHAR1. To eliminate the blanks, use the TRIM function as shown in the second SYMPUT statement.

```
data char1;
   input c $;
   call symput('char1',c);
   call symput('char2',trim(c));
   datalines;
x
;
run;
%put char1 = ***&char1***;
%put char2 = ***&char2***;
```

When this program executes, these lines are written to the SAS log:

```
   char1 = ***x       ***
   char2 = ***x***
```

Formatting Rules For Assigning Numeric Values

If *value* is a numeric variable, SYMPUT writes it using the BEST12. format. The resulting value is a 12-byte string with the value right-aligned within it. For example, this DATA step assigns the value of numeric variable X to the macro variables NUM1 and NUM2. The last CALL SYMPUT statement deletes undesired leading blanks by using the LEFT function to left-align the value before the SYMPUT routine assigns the value to NUM2.

```
data _null_;
   x=1;
   call symput('num1',x);
   call symput('num2',left(x));
   call symput('num3',trim(left(put(x,8.)))); /*preferred technique*/
run;
%put num1 = ***&num1***;
%put num2 = ***&num2***;
%put num3 = ***&num3***;
```

When this program executes, these lines are written to the SAS log:

```
   num1 = ***           1***
   num2 = ***1           ***
   num3 = ***1***
```

Comparisons

- SYMPUT assigns values produced in a DATA step to macro variables during program execution, but the SYMGET function returns values of macro variables to the program during program execution.
- SYMPUT is available in DATA step and SCL programs, but SYMPUTN is available only in SCL programs.
- SYMPUT assigns character values, but SYMPUTN assigns numeric values.

Example: Creating Macro Variables and Assigning Them Values from a Data Set

```
data dusty;
   input dept $ name $ salary @@;
   datalines;
bedding Watlee 18000    bedding Ives 16000
bedding Parker 9000     bedding George 8000
bedding Joiner 8000     carpet Keller 20000
carpet Ray 12000        carpet Jones 9000
gifts Johnston 8000     gifts Matthew 19000
kitchen White 8000      kitchen Banks 14000
kitchen Marks 9000      kitchen Cannon 15000
tv Jones 9000           tv Smith 8000
tv Rogers 15000         tv Morse 16000
;
proc means noprint;
   class dept;
   var salary;
   output out=stats sum=s_sal;
run;
data _null_;
   set stats;
   if _n_=1 then call symput('s_tot',trim(left(s_sal)));
   else call symput('s'||dept,trim(left(s_sal)));
run;
%put _user_;
```

When this program executes, this list of variables is written to the SAS log:

```
GLOBAL SCARPET 41000
GLOBAL SKITCHEN 46000
GLOBAL STV 48000
GLOBAL SGIFTS 27000
GLOBAL SBEDDING 59000
GLOBAL S_TOT 221000
```

CALL SYMPUTN Routine

In SCL programs, assigns a numeric value to a global macro variable.

Type: SCL call routine

See: "SYMGET Function" on page 234, "SYMGETN Function" on page 237, and "CALL SYMPUT Routine" on page 224

Syntax

CALL SYMPUTN('*macro-variable*', *value*);

Required Arguments

macro-variable
is the name of a global macro variable with no ampersand – note the single quotation marks. Or, it is the name of an SCL variable that contains the name of a global macro variable.

value
is the numeric value to assign, which can be a number or the name of a numeric SCL variable.

Details

The SYMPUTN routine assigns a numeric value to a global SAS macro variable. SYMPUTN assigns the value when the SCL program executes. You can also use SYMPUTN to assign the value of a macro variable whose name is stored in an SCL variable. For example, to assign the value of SCL variable UNITNUM to SCL variable UNITVAR, which contains 'UNIT', submit the following:

```
call symputn(unitvar,unitnum)
```

You must use SYMPUTN with a CALL statement.

Note: It is inefficient to use an ampersand (&) to reference a macro variable that was created with CALL SYMPUTN. Instead, use SYMGETN. It is also inefficient to use CALL SYMPUTN to store a variable that does not contain a numeric value.

Comparisons

- SYMPUTN assigns numeric values, but SYMPUT assigns character values.
- SYMPUTN is available only in SCL programs, but SYMPUT is available in DATA step programs and SCL programs.
- SYMPUTN assigns numeric values, but SYMGETN retrieves numeric values.

Example: Storing the Value 1000 in the Macro Variable UNIT When the SCL Program Executes

This statement stores the value 1000 in the macro variable UNIT when the SCL program executes:

```
call symputn('unit',1000);
```

CALL SYMPUTX Routine

Assigns a value to a macro variable, and removes both leading and trailing blanks.

Category: Macro

See: "CALL SYMPUTX Routine" in *SAS Functions and CALL Routines: Reference*

Syntax

CALL SYMPUTX(*macro-variable*, *value* <,*symbol-table*>);

Chapter 16
DATA Step Functions for Macros

DATA Step Functions for Macros

You can interact with the Macro Facility using DATA step functions.

Dictionary

RESOLVE Function

Resolves the value of a text expression during DATA step execution.

Type: DATA step function

Syntax

RESOLVE(*argument*)

Required Argument

argument
can be one of the following items:

- a text expression enclosed in single quotation marks (to prevent the macro processor from resolving the argument while the DATA step is being constructed). When a macro variable value contains a macro variable reference, RESOLVE attempts to resolve the reference. If *argument* references a nonexistent macro variable, RESOLVE returns the unresolved reference. These examples using text expressions show how to assign the text generated by macro LOCATE or assign the value of the macro variable NAME:

```
x=resolve('%locate');
x=resolve('&name');
```

- the name of a DATA step variable whose value is a text expression. For example, this example assigns the value of the text expression in the current value of the DATA step variable ADDR1 to X:

```
addr1='&locate';
x=resolve(addr1);
```

- a character expression that produces a text expression for resolution by the macro facility. For example, this example uses the current value of the DATA step variable STNUM in building the name of a macro:

```
x=resolve('%state'||left(stnum));
```

Details

The RESOLVE function returns a character value that is the maximum length of a DATA step character variable unless you specifically assign the target variable a shorter length. A returned value that is longer is truncated.

If RESOLVE cannot locate the macro variable or macro identified by the argument, it returns the argument without resolution and the macro processor issues a warning message.

You can create a macro variable with the SYMPUT routine and use RESOLVE to resolve it in the same DATA step.

Comparisons

- RESOLVE resolves the value of a text expression during execution of a DATA step or SCL program. Whereas a macro variable reference resolves when a DATA step is being constructed or an SCL program is being compiled. For this reason, the resolved value of a macro variable reference is constant during execution of a DATA step or SCL program. However, RESOLVE can return a different value for a text expression in each iteration of the program.
- RESOLVE accepts a wider variety of arguments than the SYMGET function accepts. SYMGET resolves only a single macro variable but RESOLVE resolves any macro expression. Using RESOLVE might result in the execution of macros and resolution of more than one macro variable.
- When a macro variable value contains an additional macro variable reference, RESOLVE attempts to resolve the reference, but SYMGET does not.
- If *argument* references a nonexistent macro variable, RESOLVE returns the unresolved reference, whereas SYMGET returns a missing value.
- Because of its greater flexibility, RESOLVE requires slightly more computer resources than SYMGET.

Example: Resolving Sample References

This example shows RESOLVE used with a macro variable reference, a macro invocation, and a DATA step variable whose value is a macro invocation.

```
%let event=Holiday;
%macro date;
   New Year
%mend date;
data test;
   length var1-var3 $ 15;
   when='%date';
   var1=resolve('&event'); /* macro variable reference */
   var2=resolve('%date');  /* macro invocation */
   var3=resolve(when);     /* DATA step variable with macro invocation */
   put var1=  var2=  var3=;
run;
```

When this program executes, these lines are written to the SAS log:

```
VAR1=Holiday  VAR2=New Year  VAR3=New Year
NOTE: The data set WORK.TEST has 1 observations and 4 variables.
```

SYMEXIST Function

Returns an indication of the existence of a macro variable.

Type: DATA step function

Syntax

SYMEXIST (*argument*)

Required Argument

argument
can be one of the following items:

- the name of a macro variable within quotation marks but without an ampersand
- the name of a DATA step character variable, specified with no quotation marks, which contains a macro variable name
- a character expression that constructs a macro variable name

Details

The SYMEXIST function searches any enclosing local symbol tables and then the global symbol table for the indicated macro variable. The SYMEXIST function returns one of the following values:

- `1` if the macro variable is found
- `0` if the macro variable is not found

Example: Using SYMEXIST Function

The following example of the %TEST macro contains the SYMEXIST function:

```
%global x;
      %macro test;
      %local y;
      data null;
         if symexist("x") then put "x EXISTS";
                               else put "x does not EXIST";
         if symexist("y") then put "y EXISTS";
                               else put "y does not EXIST";
         if symexist("z") then put "z EXISTS";
                               else put "z does not EXIST";
      run;
      %mend test;
      %test;
```

In the previous example, executing the %TEST macro, which contains the SYMEXIST function, writes the following output to the SAS log:

```
x EXISTS
y EXISTS
z does not EXIST
```

SYMGET Function

Returns the value of a macro variable to the DATA step during DATA step execution.

Type: DATA step function

See: "RESOLVE Function" on page 231, "SYMGETN Function" on page 237, "CALL SYMPUT Routine" on page 224, and "CALL SYMPUTN Routine" on page 228

Syntax

SYMGET(*argument*)

Required Argument

argument

can be one of the following items:

- the name of a macro variable within quotation marks but without an ampersand. When a macro variable value contains another macro variable reference, SYMGET does not attempt to resolve the reference. If *argument* references a nonexistent macro variable, SYMGET returns a missing value. This example shows how to assign the value of the macro variable G to the DATA step variable X.

  ```
  x=symget('g');
  ```

- the name of a DATA step character variable, specified with no quotation marks, which contains names of one or more macro variables. If the value is not a valid SAS name, or if the macro processor cannot find a macro variable of that name, SAS writes a note to the log that the function has an illegal argument and sets the resulting value to missing. For example, these statements assign the value stored in the DATA step variable CODE, which contains a macro variable name, to the DATA step variable KEY:

```
length key $ 8;
input code $;
key=symget(code);
```

Each time the DATA step iterates, the value of CODE supplies the name of a macro variable whose value is then assigned to KEY.

- a character expression that constructs a macro variable name. For example, this statement assigns the letter **s** and the number of the current iteration (using the automatic DATA step variable _N_).

```
score=symget('s'||left(_n_));
```

Details

SYMGET returns a character value that is the maximum length of a DATA step character variable. A returned value that is longer is truncated.

If SYMGET cannot locate the macro variable identified as the argument, it returns a missing value, and the program issues a message for an illegal argument to a function.

SYMGET can be used in all SAS language programs, including SCL programs. Because it resolves variables at program execution instead of macro execution, SYMGET should be used to return macro values to DATA step views, SQL views, and SCL programs.

Comparisons

- SYMGET returns values of macro variables during program execution, whereas the SYMPUT function assigns values that are produced by a program to macro variables during program execution.
- SYMGET accepts fewer types of arguments than the RESOLVE function. SYMGET resolves only a single macro variable. Using RESOLVE might result in the execution of macros and further resolution of values.
- SYMGET is available in all SAS programs, but SYMGETN is available only in SCL programs.

Example: Retrieving Variable Values Previously Assigned from a Data Set

```
data dusty;
   input dept $ name $ salary @@;
   datalines;
bedding Watlee 18000    bedding Ives 16000
bedding Parker 9000     bedding George 8000
bedding Joiner 8000     carpet Keller 20000
carpet Ray 12000        carpet Jones 9000
gifts Johnston 8000     gifts Matthew 19000
kitchen White 8000      kitchen Banks 14000
kitchen Marks 9000      kitchen Cannon 15000
tv Jones 9000           tv Smith 8000
tv Rogers 15000         tv Morse 16000
;
proc means noprint;
   class dept;
   var salary;
   output out=stats sum=s_sal;
```

```
run;
proc print data=stats;
   var dept s_sal;
   title "Summary of Salary Information";
   title2 "For Dusty Department Store";
run;
data _null_;
   set stats;
   if _n_=1 then call symput('s_tot',s_sal);
   else call symput('s'||dept,s_sal);
run;
data new;
   set dusty;
   pctdept=(salary/symget('s'||dept))*100;
   pcttot=(salary/&s_tot)*100;
run;
proc print data=new split="*";
   label dept    ="Department"
         name    ="Employee"
         pctdept="Percent of *Department* Salary"
         pcttot ="Percent of *   Store   * Salary";
   format pctdept pcttot 4.1;
   title  "Salary Profiles for Employees";
   title2 "of Dusty Department Store";
run;
```

This program produces the following output:

Output 16.1 *Summary of Salary Information*

Summary of Salary Information
For Dusty Department Store

Obs	dept	s_sal
1		221000
2	bedding	59000
3	carpet	41000
4	gifts	27000
5	kitchen	46000
6	tv	48000

Output 16.2 *Salary Profiles for Employees*

Salary Profiles for Employees of Dusty Department Store

Obs	Department	Employee	salary	Percent of Department Salary	Percent of Store Salary
1	bedding	Watlee	18000	30.5	8.1
2	bedding	Ives	16000	27.1	7.2
3	bedding	Parker	9000	15.3	4.1
4	bedding	George	8000	13.6	3.6
5	bedding	Joiner	8000	13.6	3.6
6	carpet	Keller	20000	48.8	9.0
7	carpet	Ray	12000	29.3	5.4
8	carpet	Jones	9000	22.0	4.1
9	gifts	Johnston	8000	29.6	3.6
10	gifts	Matthew	19000	70.4	8.6
11	kitchen	White	8000	17.4	3.6
12	kitchen	Banks	14000	30.4	6.3
13	kitchen	Marks	9000	19.6	4.1
14	kitchen	Cannon	15000	32.6	6.8
15	tv	Jones	9000	18.8	4.1
16	tv	Smith	8000	16.7	3.6
17	tv	Rogers	15000	31.3	6.8
18	tv	Morse	16000	33.3	7.2

SYMGETN Function

In SAS Component Control Language (SCL) programs, returns the value of a global macro variable as a numeric value.

Type: SCL function

See: "SYMGET Function" on page 234, "CALL SYMPUT Routine" on page 224, and "CALL SYMPUTN Routine" on page 228

Syntax

SCL-variable=***SYMGETN***('*macro-variable*');

Required Arguments

SCL variable
is the name of a numeric SCL variable to contain the value stored in *macro-variable*.

macro-variable
is the name of a global macro variable with no ampersand – note the single quotation marks. Or, the name of an SCL variable that contains the name of a global macro variable.

Details

SYMGETN returns the value of a global macro variable as a numeric value and stores it in the specified numeric SCL variable. You can also use SYMGETN to retrieve the value of a macro variable whose name is stored in an SCL variable. For example, to retrieve the value of SCL variable UNITVAR, whose value is 'UNIT', submit the following code:

```
unitnum=symgetn(unitvar)
```

SYMGETN returns values when SCL programs execute. If SYMGETN cannot locate *macro-variable*, it returns a missing value.

To return the value stored in a macro variable when an SCL program compiles, use a macro variable reference in an assignment statement:

```
SCL variable=&macro-variable;
```

Note: It is inefficient to use SYMGETN to retrieve values that are not assigned with SYMPUTN and values that are not numeric.

Comparisons

- SYMGETN is available only in SCL programs, but SYMGET is available in DATA step programs and SCL programs.
- SYMGETN retrieves values, but SYMPUTN assigns values.

Example: Storing a Macro Variable Value as a Numeric Value in an SCL Program

This statement stores the value of the macro variable UNIT in the SCL variable UNITNUM when the SCL program executes:

```
unitnum=symgetn('unit');
```

SYMGLOBL Function

Returns an indication as to whether a macro variable is global in scope to the DATA step during DATA step execution.

Type: DATA step function

Syntax

SYMGLOBL (*argument*)

Required Argument

argument
can be one of the following items:

- the name of a macro variable within quotation marks but without an ampersand
- the name of a DATA step character variable, specified with no quotation marks, that contains a macro variable name
- a character expression that constructs a macro variable name

Details

The SYMGLOBL function searches enclosing scopes for the indicated macro variable and returns a value of `1` if the macro variable is found in the global symbol table, otherwise it returns a `0`. See "Scopes of Macro Variables" on page 43 for more information about the global and local symbol tables and macro variable scopes.

Example: Using SYMGLOBL Function

The following example of the %TEST macro contains the SYMGLOBL function:

```
%global x;
      %macro test;
      %local y;
      data null;
         if symglobl("x") then put "x is GLOBAL";
                              else put "x is not GLOBAL";
         if symglobl("y") then put "y is GLOBAL";
                              else put "y is not GLOBAL";
         if symglobl("z") then put "z is GLOBAL";
                              else put "z is not GLOBAL";
      run;
      %mend test;
      %test;
```

In the previous example, executing the %TEST macro, which contains the SYMGLOBL function, writes the following output to the SAS log:

```
x is GLOBAL
y is not GLOBAL
z is not GLOBAL
```

SYMLOCAL Function

Returns an indication as to whether a macro variable is local in scope to the DATA step during DATA step execution.

Type: DATA step function

Syntax

SYMLOCAL (*argument*)

Required Argument

argument
can be one of the following items:

- the name of a macro variable within quotation marks but without an ampersand
- the name of a DATA step character variable, specified with no quotation marks, that contains a macro variable name
- a character expression that constructs a macro variable name

Details

The SYMLOCAL function searches enclosing scopes for the indicated macro variable and returns a value of `1` if the macro variable is found in a local symbol table, otherwise it returns a `0`. See “Scopes of Macro Variables” on page 43 for more information about the global and local symbol tables and macro variable scopes.

Example: Using SYMLOCAL Function

The following example of the %TEST macro contains the SYMLOCAL function:

```
%global x;
     %macro test;
     %local y;
     data null;
        if symlocal("x") then put "x is LOCAL";
                               else put "x is not LOCAL";
        if symlocal("y") then put "y is LOCAL";
                               else put "y is not LOCAL";
        if symlocal("z") then put "z is LOCAL";
                               else put "z is not LOCAL";
     run;
     %mend test;
     %test;
```

In the previous example, executing the %TEST macro, which contains the SYMLOCAL function, writes the following output to the SAS log:

```
x is not LOCAL
y is LOCAL
z is not LOCAL
```

Chapter 17
Macro Functions

Macro Functions

A macro language function processes one or more arguments and produces a result.

Dictionary

%BQUOTE and %NRBQUOTE Functions

Mask special characters and mnemonic operators in a resolved value at macro execution.

Type: Macro quoting function

See: "%QUOTE and %NRQUOTE Functions" on page 248 and "%SUPERQ Function" on page 259

Syntax

%BQUOTE (*character string* | *text expression*)

%NRBQUOTE (*character string* | *text expression*)

Details

The %BQUOTE and %NRBQUOTE functions mask a character string or resolved value of a text expression during execution of a macro or macro language statement. They mask the following special characters and mnemonic operators:

```
' " ( ) + - * / < > = ¬ ^ ~ ; , # blank
AND OR NOT EQ NE LE LT GE GT IN
```

In addition, %NRBQUOTE masks:

```
& %
```

%NRBQUOTE is most useful when the resolved value of an argument might contain

- strings that look like macro variable references but are not, so the macro processor should not attempt to resolve them when it next encounters them.
- macro invocations that you do not want the macro processor to attempt to resolve when it next encounters them.

Note: The maximum level of nesting for the macro quoting functions is 10.

Tip: You can use %BQUOTE and %NRBQUOTE for all execution-time macro quoting because they mask all characters and mnemonic operators that can be interpreted as elements of macro language.

Quotation marks (`' "`) do not have to be marked.

For a description of quoting in SAS macro language, see "Macro Quoting" on page 80.

Comparisons

%NRBQUOTE and the %SUPERQ function mask the same items. However, %SUPERQ does not attempt to resolve a macro variable reference or a macro invocation that occurs in the value of the specified macro variable. %NRBQUOTE does attempt to resolve such references.%BQUOTE and %NRBQUOTE do not require that you mark quotation marks.

Example: Quoting a Variable

This example tests whether a filename passed to the macro FILEIT starts with a quotation mark. Based on that evaluation, the macro creates the correct FILE command.

```
%macro fileit(infile);
  %if %bquote(&infile) NE %then
    %do;
        %let char1 = %bquote(%substr(&infile,1,1));
        %if %bquote(&char1) = %str(%')
            or %bquote(&char1) = %str(%")
        %then %let command=FILE &infile;
        %else %let command=FILE "&infile";
    %end;
  %put &command;
%mend fileit;
%fileit(myfile)
%fileit('myfile')
```

When this program executes, the following is written to the log:

```
FILE "myfile"
FILE 'myfile'
```

%EVAL Function

Evaluates arithmetic and logical expressions using integer arithmetic.

Type: Macro evaluation function

See: "%SYSEVALF Function" on page 263

Syntax

%EVAL (*arithmetic or logical expression*)

Details

The %EVAL function evaluates integer arithmetic or logical expressions. %EVAL operates by converting its argument from a character value to a numeric or logical expression. Then, it performs the evaluation. Finally, %EVAL converts the result back to a character value and returns that value.

If all operands can be interpreted as integers, the expression is treated as arithmetic. If at least one operand cannot be interpreted as numeric, the expression is treated as logical. If a division operation results in a fraction, the fraction is truncated to an integer.

Logical, or Boolean, expressions return a value that is evaluated as true or false. In the macro language, any numeric value other than 0 is true and a value of 0 is false.

%EVAL accepts only operands in arithmetic expressions that represent integers (in standard or hexadecimal form). Operands that contain a period character cause an error when they are part of an integer arithmetic expression. The following examples show correct and incorrect usage, respectively:

```
%let d=%eval(10+20);        /* Correct usage   */
%let d=%eval(10.0+20.0);    /* Incorrect usage */
```

Because %EVAL does not convert a value containing a period to a number, the operands are evaluated as character operands. When %EVAL encounters a value containing a period, it displays an error message about finding a character operand where a numeric operand is required.

An expression that compares character values in the %EVAL function uses the sort sequence of the operating environment for the comparison. Refer to "The SORT PROCEDURE" in the *Base SAS Procedures Guide* for more information about operating environment sort sequences.

All parts of the macro language that evaluate expressions (for example, %IF and %DO statements) call %EVAL to evaluate the condition. For a complete discussion of how macro expressions are evaluated, see "Macro Expressions" on page 71.

Comparisons

%EVAL performs integer evaluations, but %SYSEVALF performs floating point evaluations.

Examples

Example 1: Illustrating Integer Arithmetic Evaluation

These statements illustrate different types of evaluations:

```
%let a=1+2;
%let b=10*3;
%let c=5/3;
%let eval_a=%eval(&a);
%let eval_b=%eval(&b);
%let eval_c=%eval(&c);
%put &a is &eval_a;
%put &b is &eval_b;
%put &c is &eval_c;
```

When these statements are submitted, the following is written to the SAS log:

```
1+2 is 3
10*3 is 30
5/3 is 1
```

The third %PUT statement shows that %EVAL discards the fractional part when it performs division on integers that would result in a fraction:

Example 2: Incrementing a Counter

The macro TEST uses %EVAL to increment the value of the macro variable I by 1. Also, the %DO %WHILE statement calls %EVAL to evaluate whether I is greater than the value of the macro variable FINISH.

```
%macro test(finish);
   %let i=1;
   %do %while (&i<&finish);
      %put the value of i is &i;
      %let i=%eval(&i+1);
   %end;
```

```
%mend test;
%test(5)
```

When this program executes, these lines are written to the SAS log:

```
The value of i is 1
The value of i is 2
The value of i is 3
The value of i is 4
```

Example 3: Evaluating Logical Expressions

Macro COMPARE compares two numbers.

```
%macro compare(first,second);
   %if &first>&second %then %put &first > &second;
   %else %if &first=&second %then %put &first = &second;
   %else %put &first<&second;
%mend compare;
%compare(1,2)
%compare(-1,0)
```

When this program executes, these lines are written to the SAS log:

```
1 < 2
-1 < 0
```

%INDEX Function

Returns the position of the first character of a string.

Type: Macro function

Syntax

%INDEX (*source, string*)

Required Arguments

source
is a character string or text expression.

string
is a character string or text expression.

Details

The %INDEX function searches *source* for the first occurrence of *string* and returns the position of its first character. If *string* is not found, the function returns 0.

Example: Locating a Character

The following statements find the first character **V** in a string:

```
%let a=a very long value;
%let b=%index(&a,v);
%put V appears at position &b..;
```

When these statements execute, the following line is written to the SAS log:

```
V appears at position 3.
```

%LENGTH Function

Returns the length of a string.

Type: Macro function

Syntax

%LENGTH (*character string* | *text expression*)

Details

If the argument is a character string, %LENGTH returns the length of the string. If the argument is a text expression, %LENGTH returns the length of the resolved value. If the argument has a null value, %LENGTH returns 0.

Example: Returning String Lengths

The following statements find the lengths of character strings and text expressions.

```
%let a=Happy;
%let b=Birthday;
%put The length of &a is %length(&a).;
%put The length of &b is %length(&b).;
%put The length of &a &b To You is %length(&a &b to you).;
```

When these statements execute, the following is written to the SAS log:

```
The length of Happy is 5.
The length of Birthday is 8.
The length of Happy Birthday To You is 21.
```

%NRBQUOTE Function

Masks special characters, including & and %, and mnemonic operators in a resolved value at macro execution.

Type: Macro quoting function

See: "%BQUOTE and %NRBQUOTE Functions" on page 242

Syntax

%NRBQUOTE (*character string* | *text expression*)

Without Arguments

Note that the maximum level of nesting for the macro quoting functions is 10.

%NRQUOTE Function

Masks special characters, including & and %, and mnemonic operators in a resolved value at macro execution.

Type: Macro quoting function

See: "%QUOTE and %NRQUOTE Functions" on page 248

Syntax

%NRQUOTE (*character string* | *text expression*)

Without Arguments

Note that the maximum level of nesting for the macro quoting functions is 10.

%NRSTR Function

Masks special characters, including & and %, and mnemonic operators in constant text during macro compilation.

Type: Macro quoting function

See: "%STR and %NRSTR Functions" on page 254

Syntax

%NRSTR (*character-string*)

Without Arguments

Note that the maximum level of nesting for the macro quoting functions is 10.

%QSCAN Function

Searches for a word and masks special characters and mnemonic operators.

Type: Macro function

Syntax

%QSCAN (*argument*,*n*<,charlist<,modifiers>>)

Without Arguments

"%SCAN and %QSCAN Functions" on page 250

%QSUBSTR Function

Produces a substring and masks special characters and mnemonic operators.

Type: Macro function

Syntax

%QSUBSTR (*argument, position*<, *length*>)

Without Arguments

See “%SUBSTR and %QSUBSTR Functions” on page 257

%QSYSFUNC Function

Executes functions and masks special characters and mnemonic operators.

Type: Macro function

Syntax

%QSYSFUNC (*function*(*argument-1* <...*argument-n*>)<, *format*>)

Without Arguments

See“%SYSFUNC and %QSYSFUNC Functions” on page 268

%QUOTE and %NRQUOTE Functions

Mask special characters and mnemonic operators in a resolved value at macro execution.

Type: Macro quoting function

See: “%BQUOTE and %NRBQUOTE Functions” on page 242, “%NRBQUOTE Function” on page 246, “%NRSTR Function” on page 247, and “%SUPERQ Function” on page 259

Syntax

%QUOTE (*character string* | *text expression*)

%NRQUOTE (*character string* | *text expression*)

Details

The %QUOTE and %NRQUOTE functions mask a character string or resolved value of a text expression during execution of a macro or macro language statement. They mask the following special characters and mnemonic operators:

```
+ - * / < > = ¬ ^ ~ ; , # blank
AND OR NOT EQ NE LE LT GE GT IN
```

They also mask the following characters when they occur in pairs and when they are not matched and are marked by a preceding `%`

```
' "
```

In addition, %NRQUOTE masks

```
& %
```

%NRQUOTE is most useful when an argument might contain a macro variable reference or macro invocation that you do not want resolved.

For a description of quoting in SAS macro language, see “Macro Quoting” on page 80.

Note that the maximum level of nesting for the macro quoting functions is 10.

Comparisons

- %QUOTE and %NRQUOTE mask the same items as %STR and %NRSTR, respectively. However, %STR and %NRSTR mask constant text instead of a resolved value. And, %STR and %NRSTR work when a macro compiles, while %QUOTE and %NRQUOTE work when a macro executes.
- The %BQUOTE and %NRBQUOTE functions do not require that quotation marks without a match be marked with a preceding `%`, while %QUOTE and %NRQUOTE do.
- %QUOTE and %NRQUOTE mask resolved values, while the %SUPERQ function prevents resolution of any macro invocations or macro variable references that might occur in a value.

Example: Quoting a Value that Might Contain a Mnemonic Operator

The macro DEPT1 receives abbreviations for states and therefore might receive the value OR for Oregon.

```
%macro dept1(state);
      /* without %quote -- problems might occur */
   %if &state=nc %then
       %put North Carolina Department of Revenue;
   %else %put Department of Revenue;
%mend dept1;
%dept1(or)
```

When the macro DEPT1 executes, the %IF condition executes a %EVAL function, which evaluates **`or`** as a logical operator in this expression. Then the macro processor produces an error message for an invalid operand in the expression **`or=nc`**.

The macro DEPT2 uses the %QUOTE function to treat characters that result from resolving &STATE as text:

```
%macro dept2(state);
      /* with %quote function--problems are prevented */
   %if %quote(&state)=nc %then
       %put North Carolina Department of Revenue;
   %else %put Department of Revenue;
%mend dept2;
%dept2(or)
```

The %IF condition now compares the strings **`or`** and **`nc`** and writes to the SAS log:

```
Department of Revenue
```

%QUPCASE Function

Converts a value to uppercase and returns a result that masks special characters and mnemonic operators.

Type: Macro function

Syntax

%QUPCASE (*character string* | *text expression*)

Without Arguments

See "%UPCASE and %QUPCASE Functions" on page 275

%SCAN and %QSCAN Functions

Search for a word that is specified by its position in a string.

Type: Macro function

See: "%NRBQUOTE Function" on page 246 and "%STR and %NRSTR Functions" on page 254

Syntax

%SCAN(*argument*, *n*<,*charlist*<,*modifiers*>>)

%QSCAN(*argument*, *n*<,*charlist*<,*modifiers*>>)

Required Arguments

argument
is a character string or a text expression. If *argument* might contain a special character or mnemonic operator, listed below, use %QSCAN. If *argument* contains a comma, enclose *argument* in a quoting function such as %BQUOTE(*argument*).

n
is an integer or a text expression that yields an integer, which specifies the position of the word to return. (An implied %EVAL gives *n* numeric properties.) If *n* is greater than the number of words in *argument*, the functions return a null string.

Note: When you are using Version 8 or greater, if *n* is negative, %SCAN examines the character string and selects the word that starts at the end of the string and searches backward.

charlist
specifies an optional character expression that initializes a list of characters. This list determines which characters are used as the delimiters that separate words. The following rules apply:

- By default, all characters in *charlist* are used as delimiters.

- If you specify the K modifier in the *modifier* argument, then all characters that are not in *charlist* are used as delimiters.

Tip: You can add more characters to *charlist* by using other modifiers.

modifier

specifies a character constant, a variable, or an expression in which each non-blank character modifies the action of the %SCAN function. Blanks are ignored. You can use the following characters as modifiers:

a or A	adds alphabetic characters to the list of characters.
b or B	scans backward from right to left instead of from left to right, regardless of the sign of the *count* argument.
c or C	adds control characters to the list of characters.
d or D	adds digits to the list of characters.
f or F	adds an underscore and English letters (that is, valid first characters in a SAS variable name using VALIDVARNAME=V7) to the list of characters.
g or G	adds graphic characters to the list of characters. Graphic characters are characters that, when printed, produce an image on paper.
h or H	adds a horizontal tab to the list of characters.
i or I	ignores the case of the characters.
k or K	causes all characters that are not in the list of characters to be treated as delimiters. That is, if K is specified, then characters that are in the list of characters are kept in the returned value rather than being omitted because they are delimiters. If K is not specified, then all characters that are in the list of characters are treated as delimiters.
l or L	adds lowercase letters to the list of characters.
m or M	specifies that multiple consecutive delimiters, and delimiters at the beginning or end of the *string* argument, refer to words that have a length of zero. If the M modifier is not specified, then multiple consecutive delimiters are treated as one delimiter, and delimiters at the beginning or end of the *string* argument are ignored.
n or N	adds digits, an underscore, and English letters (that is, the characters that can appear in a SAS variable name using VALIDVARNAME=V7) to the list of characters.
o or O	processes the *charlist* and *modifier* arguments only once, rather than every time the %SCAN function is called. Using the O modifier in the DATA step (excluding WHERE clauses), or in the SQL procedure can make %SCAN run faster when you call it in a loop where the *charlist* and *modifier* arguments do not change. The O modifier applies separately to each instance of the %SCAN function in your SAS code, and does not cause all instances of the %SCAN function to use the same delimiters and modifiers.
p or P	adds punctuation marks to the list of characters.
q or Q	ignores delimiters that are inside of substrings that are enclosed in quotation marks. If the value of the *string* argument contains unmatched quotation marks, then scanning from left to right will produce different words than scanning from right to left.

r or R — removes leading and trailing blanks from the word that %SCAN returns. If you specify both the Q and R modifiers, then the %SCAN function first removes leading and trailing blanks from the word. Then, if the word begins with a quotation mark, %SCAN also removes one layer of quotation marks from the word.

s or S — adds space characters to the list of characters (blank, horizontal tab, vertical tab, carriage return, line feed, and form feed).

t or T — trims trailing blanks from the *string* and *charlist* arguments. If you want to remove trailing blanks from only one character argument instead of both character arguments, then use the TRIM function instead of the %SCAN function with the T modifier.

u or U — adds uppercase letters to the list of characters.

w or W — adds printable (writable) characters to the list of characters.

x or X — adds hexadecimal characters to the list of characters.

Tip: If the *modifier* argument is a character constant, then enclose it in quotation marks. Specify multiple modifiers in a single set of quotation marks. A *modifier* argument can also be expressed as a character variable or expression.

Details

The %SCAN and %QSCAN functions search *argument* and return the *n*th word. A word is one or more characters separated by one or more delimiters.

%SCAN does not mask special characters or mnemonic operators in its result, even when the argument was previously masked by a macro quoting function. %QSCAN masks the following special characters and mnemonic operators in its result:

```
& % ' " ( ) + - * / < > = ¬ ^ ~ ; , # blank
AND OR NOT EQ NE LE LT GE GT IN
```

Definition of "Delimiter" and "Word"

A delimiter is any of several characters that are used to separate words. You can specify the delimiters in the *charlist* and *modifier* arguments.

If you specify the Q modifier, then delimiters inside of substrings that are enclosed in quotation marks are ignored.

In the %SCAN function, "word" refers to a substring that has all of the following characteristics:

- is bounded on the left by a delimiter or the beginning of the string
- is bounded on the right by a delimiter or the end of the string
- contains no delimiters

A word can have a length of zero if there are delimiters at the beginning or end of the string, or if the string contains two or more consecutive delimiters. However, the %SCAN function ignores words that have a length of zero unless you specify the M modifier.

Using Default Delimiters in ASCII and EBCDIC Environments

If you use the %SCAN function with only two arguments, then the default delimiters depend on whether your computer uses ASCII or EBCDIC characters.

- If your computer uses ASCII characters, then the default delimiters are as follows:

blank ! $ % & ( ) * + , - . / ; < ^¦

In ASCII environments that do not contain the ^ character, the %SCAN function uses the ~ character instead.

- If your computer uses EBCDIC characters, then the default delimiters are as follows:

  blank ! $ % & ( ) * + , - . / ; < ¬ | ¢¦

If you use the *modifier* argument without specifying any characters as delimiters, then the only delimiters that will be used are delimiters that are defined by the *modifier* argument. In this case, the lists of default delimiters for ASCII and EBCDIC environments are not used. In other words, modifiers add to the list of delimiters that are specified by the *charlist* argument. Modifiers do not add to the list of default modifiers.

Using the %SCAN Function with the M Modifier

If you specify the M modifier, then the number of words in a string is defined as one plus the number of delimiters in the string. However, if you specify the Q modifier, delimiters that are inside quotation marks are ignored.

If you specify the M modifier, then the %SCAN function returns a word with a length of zero if one of the following conditions is true:

- The string begins with a delimiter and you request the first word.
- The string ends with a delimiter and you request the last word.
- The string contains two consecutive delimiters and you request the word that is between the two delimiters.

Using the %SCAN Function without the M Modifier

If you do not specify the M modifier, then the number of words in a string is defined as the number of maximal substrings of consecutive non-delimiters. However, if you specify the Q modifier, delimiters that are inside quotation marks are ignored.

If you do not specify the M modifier, then the %SCAN function does the following:

- ignores delimiters at the beginning or end of the string
- treats two or more consecutive delimiters as if they were a single delimiter

If the string contains no characters other than delimiters, or if you specify a count that is greater in absolute value than the number of words in the string, then the %SCAN function returns one of the following:

- a single blank when you call the %SCAN function from a DATA step
- a string with a length of zero when you call the %SCAN function from the macro processor

Using Null Arguments

The %SCAN function allows character arguments to be null. Null arguments are treated as character strings with a length of zero. Numeric arguments cannot be null.

Comparisons

%QSCAN masks the same characters as the %NRBQUOTE function.

Example: Comparing the Actions of %SCAN and %QSCAN

This example illustrates the actions of %SCAN and %QSCAN.

```
%macro a;
   aaaaaa
%mend a;
%macro b;
   bbbbbb
%mend b;
%macro c;
   cccccc
%mend c;
%let x=%nrstr(%a*%b*%c);
%put X: &x;
%put The third word in X, with SCAN: %scan(&x,3,*);
%put The third word in X, with QSCAN: %qscan(&x,3,*);
```

The %PUT statement writes these lines to the log:

```
X: %a*%b*%c
The third word in X, with SCAN: cccccc
The third word in X, with QSCAN: %c
```

%STR and %NRSTR Functions

Mask special characters and mnemonic operators in constant text at macro compilation.

Type: Macro quoting function

See: "%NRQUOTE Function" on page 247

Syntax

%STR (*character-string*)

%NRSTR (*character-string*)

Details

The %STR and %NRSTR functions mask a character string during compilation of a macro or macro language statement. They mask the following special characters and mnemonic operators:

```
+ - * / < > = ¬ ^ ~ ; ,  # blank
AND OR NOT EQ NE LE LT GE GT IN
```

They also mask the following characters when they occur in pairs and when they are not matched and are marked by a preceding %

```
' " ( )
```

In addition, %NRSTR also masks the following characters:

```
& %
```

Table 17.1 *Using %STR and %NSTR Arguments*

Argument	Use
Percent sign before a quotation mark - for example, %' or %",	Percent sign with quotation mark EXAMPLE: %let percent=%str(Jim%'s office);
Percent sign before a parenthesis - for example, %(or %)	Two percent signs (%%): EXAMPLE: %let x=%str(20%%);
Character string with the comment symbols /* or -->	%STR with each character EXAMPLE: %str(/) %str(*) *comment-text* %str(*)%str(/)

%STR is most useful for character strings that contain

- a semicolon that should be treated as text rather than as part of a macro program statement
- blanks that are significant
- a quotation mark or parenthesis without a match

Putting the same argument within nested %STR and %QUOTE functions is redundant. This example shows an argument that is masked at macro compilation by the %STR function and so remains masked at macro execution. Thus, in this example, the %QUOTE function used here has no effect.

```
%quote(%str(argument))
```

CAUTION:

Do not use %STR to enclose other macro functions or macro invocations that have a list of parameter values. Because %STR masks parentheses without a match, the macro processor does not recognize the arguments of a function or the parameter values of a macro invocation.

For a description of quoting in SAS macro language, see “Macro Quoting” on page 80.

Note: The maximum level of nesting for macro quoting functions is 10.

Comparisons

- Of all the macro quoting functions, only %NRSTR and %STR take effect during compilation. The other macro quoting functions take effect when a macro executes.
- %STR and %NRSTR mask the same items as %QUOTE and %NRQUOTE. However, %QUOTE and %NRQUOTE work during macro execution.
- If resolution of a macro expression produces items that need to be masked, use the %BQUOTE or %NRBQUOTE function instead of the %STR or %NRSTR function.

Examples

Example 1: Maintaining Leading Blanks

This example enables the value of the macro variable TIME to contain leading blanks.

```
%let time=%str(    now);
%put Text followed by the value of time:&time;
```

When this example is executed, these lines are written to the SAS log:

```
Text followed by the value of time:    now
```

Example 2: Protecting a Blank So That It Will Be Compiled as Text

This example specifies that %QSCAN use a blank as the delimiter between words.

```
%macro words(string);
   %local count word;
   %let count=1;
   %let word=%qscan(&string,&count,%str( ));
   %do %while(&word ne);
      %let count=%eval(&count+1);
      %let word=%qscan(&string,&count,%str( ));
   %end;
   %let count=%eval(&count-1);
   %put The string contains &count words.;
%mend words;
%words(This is a very long string)
```

When this program executes, these lines are written to the SAS log:

```
The string contains 6 words.
```

Example 3: Quoting a Value That Might Contain a Macro Reference

The macro REVRS reverses the characters produced by the macro TEST. %NRSTR in the %PUT statement protects **%test&test** so that it is compiled as text and not interpreted.

```
%macro revrs(string);
   %local nstring;
   %do i=%length(&string) %to 1 %by -1;
      %let nstring=&nstring%qsubstr(&string,&i,1);
   %end;
 &nstring
%mend revrs;
%macro test;
   Two words
%mend test;
%put %nrstr(%test%test) - %revrs(%test%test);
```

When this program executes, the following lines are written to the SAS log:

```
1          %macro revrs(string);
2             %local nstring;
3             %do i=%length(&string) %to 1 %by -1;
4                %let nstring=&nstring%qsubstr(&string,&i,1);
5             %end;&nstring
6          %mend revrs;
7
8          %macro test;
9             Two words
10         %mend test;
11
12         %put %nrstr(%test%test) - %revrs(%test%test);
%test%test - sdrow owTsdrow owT
```

```
NOTE: SAS Institute Inc., SAS Campus Drive, Cary, NC USA 27513-2414
NOTE: The SAS System used:
      real time           0.28 seconds
      cpu time            0.12 seconds
```

%SUBSTR and %QSUBSTR Functions

Produce a substring of a character string.

Type: Macro function

See: "%NRBQUOTE Function" on page 246

Syntax

%SUBSTR (*argument, position*<, *length*>)

%QSUBSTR (*argument, position*<, *length*>)

Required Arguments

argument
is a character string or a text expression. If *argument* might contain a special character or mnemonic operator, listed below, use %QSUBSTR.

position
is an integer or an expression (text, logical, or arithmetic) that yields an integer, which specifies the position of the first character in the substring. If *position* is greater than the number of characters in the string, %SUBSTR and %QSUBSTR issue a warning message and return a null value. An automatic call to %EVAL causes *n* to be treated as a numeric value.

length
is an optional integer or an expression (text, logical, or arithmetic) that yields an integer that specifies the number of characters in the substring. If *length* is greater than the number of characters following *position* in *argument*, %SUBSTR and %QSUBSTR issue a warning message and return a substring containing the characters from *position* to the end of the string. By default, %SUBSTR and %QSUBSTR produce a string containing the characters from *position* to the end of the character string.

Details

The %SUBSTR and %QSUBSTR functions produce a substring of *argument*, beginning at *position*, for *length* number of characters.

%SUBSTR does not mask special characters or mnemonic operators in its result, even when the argument was previously masked by a macro quoting function. %QSUBSTR masks the following special characters and mnemonic operators:

```
& % ' " ( ) + - * / < > = ¬ ^ ~ ; , #  blank
AND OR NOT EQ NE LE LT GE GT IN
```

Comparisons

%QSUBSTR masks the same characters as the %NRBQUOTE function.

Examples

Example 1: Limiting a Fileref to Eight Characters

The macro MAKEFREF uses %SUBSTR to assign the first eight characters of a parameter as a fileref, in case a user assigns one that is longer.

```
%macro makefref(fileref,file);
   %if %length(&fileref) gt 8 %then
       %let fileref = %substr(&fileref,1,8);
   filename &fileref "&file";
%mend makefref;
%makefref(humanresource,/dept/humanresource/report96)
```

SAS sees the following statement:

```
FILENAME HUMANRES "/dept/humanresource/report96";
```

Example 2: Storing a Long Macro Variable Value in Segments

The macro SEPMSG separates the value of the macro variable MSG into 40-character units and stores each unit in a separate variable.

```
%macro sepmsg(msg);
   %let i=1;
   %let start=1;
   %if %length(&msg)>40 %then
      %do;
          %do %until(%length(&&msg&i)<40);
             %let msg&i=%qsubstr(&msg,&start,40);
             %put Message &i is: &&msg&i;
             %let i=%eval(&i+1);
             %let start=%eval(&start+40);
             %let msg&i=%qsubstr(&msg,&start);
          %end;
          %put Message &i is: &&msg&i;
      %end;
   %else %put No subdivision was needed.;
%mend sepmsg;
%sepmsg(%nrstr(A character operand was found in the %EVAL function
or %IF condition where a numeric operand is required.  A character
operand was found in the %EVAL function or %IF condition where a
numeric operand is required.));
```

When this program executes, these lines are written to the SAS log:

```
Message 1 is: A character operand was found in the %EV
Message 2 is: AL function or  %IF condition where a nu
Message 3 is: meric operand is required.  A character
Message 4 is: operand was  found in the %EVAL function
Message 5 is:  or %IF condition where a numeric operan
Message 6 is: d is required.
```

Example 3: Comparing Actions of %SUBSTR and %QSUBSTR

Because the value of C is masked by %NRSTR, the value is not resolved at compilation. %SUBSTR produces a resolved result because it does not mask special characters and mnemonic operators in C before processing it, even though the value of C had previously been masked with the %NRSTR function.

```
%let a=one;
%let b=two;
%let c=%nrstr(&a &b);
%put C: &c;
%put With SUBSTR: %substr(&c,1,2);
%put With QSUBSTR: %qsubstr(&c,1,2);
```

When these statements execute, these lines are written to the SAS log:

```
C: &a &b
With SUBSTR: one
With QSUBSTR: &a
```

%SUPERQ Function

Masks all special characters and mnemonic operators at macro execution but prevents further resolution of the value.

Type: Macro quoting function

See: "%NRBQUOTE Function" on page 246 and "%BQUOTE and %NRBQUOTE Functions" on page 242

Syntax

%SUPERQ (*argument*)

Required Argument

argument
is the name of a macro variable with no leading ampersand or a text expression that produces the name of a macro variable with no leading ampersand.

Details

The %SUPERQ function returns the value of a macro variable without attempting to resolve any macros or macro variable references in the value. %SUPERQ masks the following special characters and mnemonic operators:

```
& % ' " ( ) + - * / < > = ¬ ^ ~ ; , #  blank
AND OR NOT EQ NE LE LT GE GT IN
```

%SUPERQ is particularly useful for masking macro variables that might contain an ampersand or a percent sign when they are used with the %INPUT or %WINDOW statement, or the SYMPUT routine.

For a description of quoting in SAS macro language, see "Macro Quoting" on page 80.

Note: The maximum level of nesting for the macro quoting functions is 10.

Comparisons

- %SUPERQ is the only quoting function that prevents the resolution of macro variables and macro references in the value of the specified macro variable.
- %SUPERQ accepts only the name of a macro variable as its argument, without an ampersand, while the other quoting functions accept any text expression, including constant text, as an argument.

- %SUPERQ masks the same characters as the %NRBQUOTE function. However, %SUPERQ does not attempt to resolve anything in the value of a macro variable. %NRBQUOTE attempts to resolve any macro references or macro variable values in the argument before masking the result.

Example: Passing Unresolved Macro Variable Values

In this example, %SUPERQ prevents the macro processor from attempting to resolve macro references in the values of MV1 and MV2 before assigning them to macro variables TESTMV1 and TESTMV2.

```
data _null_;
   call symput('mv1','Smith&Jones');
   call symput('mv2','%macro abc;');
run;
%let testmv1=%superq(mv1);
%let testmv2=%superq(mv2);
%put Macro variable TESTMV1 is &testmv1;
%put Macro variable TESTMV2 is &testmv2;
```

When this program executes, these lines are written to the SAS log:

```
Macro variable TESTMV1 is Smith&Jones
Macro variable TESTMV2 is %macro abc;
```

You might think of the values of TESTMV1 and TESTMV2 as "pictures" of the original values of MV1 and MV2. The %PUT statement then writes the pictures in its text. The macro processor does not attempt resolution. It does not issue a warning message for the unresolved reference **&JONES** or an error message for beginning a macro definition inside a %LET statement.

%SYMEXIST Function

Returns an indication of the existence of a macro variable.

Type: Macro function

Syntax

%SYMEXIST(*macro-variable-name*)

Required Argument

macro-variable-name
is the name of a macro variable or a text expression that yields the name of a macro variable.

Details

The %SYMEXIST function searches any enclosing local symbol tables and then the global symbol table for the indicated macro variable and returns one of the following values:

- 1 if the macro variable is found
- 0 if the macro variable is not found

Example: Using %SYMEXIST Macro Function

The following example uses the %IF %THEN %ELSE macro statement to change the value of `1` and `0` to **`TRUE`** and **`FALSE`** respectively:

```
%global x;
%macro test;
    %local y;
        %if %symexist(x) %then %put %nrstr(%symexist(x)) = TRUE;
                         %else %put %nrstr(%symexist(x)) = FALSE;
        %if %symexist(y) %then %put %nrstr(%symexist(y)) = TRUE;
                         %else %put %nrstr(%symexist(y)) = FALSE;
        %if %symexist(z) %then %put %nrstr(%symexist(z)) = TRUE;
                         %else %put %nrstr(%symexist(z)) = FALSE;
%mend test;
%test;
```

In the previous example, executing the %TEST macro writes the following output to the SAS log:

```
%symexist(x) = TRUE
%symexist(y) = TRUE
%symexist(z) = FALSE
```

%SYMGLOBL Function

Returns an indication as to whether a macro variable is global in scope.

Type: Macro function

Syntax

%SYMGLOBL(*macro-variable-name*)

Required Argument

macro-variable-name
is a name of a macro variable or a text expression that yields the name of a macro variable.

Details

The %SYMGLOBL function searches enclosing scopes for the indicated macro variable and returns a value of `1` if the macro variable is found in the global symbol table, otherwise it returns a `0`. See “Scopes of Macro Variables” on page 43 for more information about the global and local symbol tables and macro variable scopes.

Example: Using %SYMGLOBL Macro Function

The following example uses the %IF %THEN %ELSE macro statement to change the values of `1` and `0` to **`TRUE`** and **`FALSE`** respectively:

```
%global x;
    %macro test;
        %local y;
        %if %symglobl(x) %then %put %nrstr(%symglobl(x)) = TRUE;
                         %else %put %nrstr(%symglobl(x)) = FALSE;
        %if %symglobl(y) %then %put %nrstr(%symglobl(y)) = TRUE;
                         %else %put %nrstr(%symglobl(y)) = FALSE;
        %if %symglobl(z) %then %put %nrstr(%symglobl(z)) = TRUE;
                         %else %put %nrstr(%symglobl(z)) = FALSE;
%mend test;
%test;
```

In the example above, executing the %TEST macro writes the following output to the SAS log:

```
%symglobl(x) = TRUE
%symglobl(y) = FALSE
%symglobl(z) = FALSE
```

%SYMLOCAL Function

Returns an indication as to whether a macro variable is local in scope.

Type: Macro function

Syntax

%SYMLOCAL(*macro-variable-name*)

Required Argument

macro-variable-name
is the name of a macro variable or a text expression that yields the name of a macro variable.

Details

The %SYMLOCAL searches enclosing scopes for the indicated macro variable and returns a value of `1` if the macro variable is found in a local symbol table, otherwise it returns a `0`. See "Scopes of Macro Variables" on page 43 for more information about the global and local symbol tables and macro variable scopes.

Example: Using %SYMLOCAL Macro Function

The following example uses the %IF %THEN %ELSE macro statement to change the values of `1` and `0` to **`TRUE`** and **`FALSE`** respectively:

```
%global x;
%macro test;
    %local y;
        %if %symlocal(x) %then %put %nrstr(%symlocal(x)) = TRUE;
                         %else %put %nrstr(%symlocal(x)) = FALSE;
        %if %symlocal(y) %then %put %nrstr(%symlocal(y)) = TRUE;
```

```
                                  %else %put %nrstr(%symlocal(y)) = FALSE;
         %if %symlocal(z) %then %put %nrstr(%symlocal(z)) = TRUE;
                                  %else %put %nrstr(%symlocal(z)) = FALSE;
%mend test;
%test;
```

In the example above, executing the %TEST macro writes the following output to the SAS log:

```
%symlocal(x) = FALSE
%symlocal(y) = TRUE
%symlocal(z) = FALSE
```

%SYSEVALF Function

Evaluates arithmetic and logical expressions using floating-point arithmetic.

Type: Macro function

See: "%EVAL Function" on page 243

Syntax

%SYSEVALF(*expression*<, *conversion-type*>)

Required Arguments

expression
is an arithmetic or logical expression to evaluate.

conversion-type
converts the value returned by %SYSEVALF to the type of value specified. The value can then be used in other expressions that require a value of that type. *Conversion-type* can be one of the following:

BOOLEAN
returns

- 0 if the result of the expression is 0 or missing
- 1 if the result is any other value.

Here is an example:

```
%sysevalf(1/3,boolean)      /* returns 1 */
%sysevalf(10+.,boolean)     /* returns 0 */
```

CEIL
returns a character value representing the smallest integer that is greater than or equal to the result of the expression. If the result is within 10^{-12} of an integer, the function returns a character value representing that integer. An expression containing a missing value returns a missing value along with a message noting that fact:

```
%sysevalf(1 + 1.1,ceil)       /* returns  3 */
%sysevalf(-1 -2.4,ceil)       /* returns -3 */
```

```
%sysevalf(-1 + 1.e-11,ceil)    /* returns  0 */
%sysevalf(10+.)                /* returns  . */
```

FLOOR

returns a character value representing the largest integer that is less than or equal to the result of the expression. If the result is within 10^{-12} of an integer, the function returns that integer. An expression with a missing value produces a missing value:

```
%sysevalf(-2.4,floor)          /* returns -3 */
%sysevalf(3,floor)             /* returns  3 */
%sysevalf(1.-1.e-13,floor)     /* returns  1 */
%sysevalf(.,floor)             /* returns  . */
```

INTEGER

returns a character value representing the integer portion of the result (truncates the decimal portion). If the result of the expression is within 10^{-12} of an integer, the function produces a character value representing that integer. If the result of the expression is positive, INTEGER returns the same result as FLOOR. If the result of the expression is negative, INTEGER returns the same result as CEIL. An expression with a missing value produces a missing value:

```
%put %sysevalf(2.1,integer);         /* returns  2 */
%put %sysevalf(-2.4,integer);        /* returns -2 */
%put %sysevalf(3,integer);           /* returns  3 */
%put %sysevalf(-1.6,integer);        /* returns -1 */
%put %sysevalf(1.-1.e-13,integer);   /* returns  1 */
```

Details

The %SYSEVALF function performs floating-point arithmetic and returns a value that is formatted using the BEST32. format. The result of the evaluation is always text. %SYSEVALF is the only macro function that can evaluate logical expressions that contain floating-point or missing values. Specify a conversion type to prevent problems when %SYSEVALF returns one of the following:

- missing or floating-point values to macro expressions
- macro variables that are used in other macro expressions that require an integer value

If the argument to the %SYSEVALF function contains no operator and no conversion type is specified, then the argument is returned unchanged.

For details about evaluation of expressions by the SAS macro language, see “Macro Expressions” on page 71.

Comparisons

- %SYSEVALF supports floating-point numbers. However, %EVAL performs only integer arithmetic.
- You must use the %SYSEVALF macro function in macros to evaluate floating-point expressions. However, %EVAL is used automatically by the macro processor to evaluate macro expressions.

Example: Illustrating Floating-Point Evaluation

The macro FIGUREIT performs all types of conversions for SYSEVALF values.

```
%macro figureit(a,b);
   %let y=%sysevalf(&a+&b);
```

```
      %put The result with SYSEVALF is: &y;
      %put  The BOOLEAN value is: %sysevalf(&a +&b, boolean);
      %put  The CEIL value is: %sysevalf(&a +&b, ceil);
      %put  The FLOOR value is: %sysevalf(&a +&b, floor);
      %put  The INTEGER value is: %sysevalf(&a +&b, int);
   %mend figureit;
   %figureit(100,1.597)
```

When this program executes, these lines are written to the SAS log:

```
The result with SYSEVALF is: 101.597
The BOOLEAN value is: 1
The CEIL value is: 102
The FLOOR value is: 101
The INTEGER value is: 101
```

%SYSMACEXEC Function

Returns an indication of the execution status of a macro.

Type: Macro function

Syntax

%SYSMACEXEC(*macro_name*)

Required Argument

macro_name
the name of a macro or a text expression that yields the name of the macro.

Details

The %SYSMACEXEC function returns the number 1 if the macro is currently executing. Otherwise, if the macro is not executing, the number 0 is returned.

%SYSMACEXIST Function

Returns an indication of the existence of a macro definition in the WORK.SASMACR catalog. Otherwise, the returned value is 0.

Type: Macro function

Syntax

%SYSMACEXIST(*macro-name*)

Required Argument

macro-name
the name of a macro or a text expression that yields the name of a macro.

Details

The %SYSMACEXIST function returns the number 1 if a definition for the macro exists in the WORK.SASMACR catalog. If there is not a macro definition, the returned value is 0.

%SYSMEXECDEPTH Function

Returns the nesting depth of macro execution from the point of the call to %SYSMEXECDEPTH.

Type: Macro Function

Tip: %SYSMEXECDEPTH and %SYSMEXECNAME were implemented to be used together, but it is not required.

See: %SYSMEXECNAME Function

Syntax

%SYSMEXECDEPTH

Details

To retrieve the nesting level of the currently executing macro, use the %SYSMEXECDEPTH. This function returns a number indicating the depth of the macro in nested macro calls. The following are the %SYSMEXECDEPTH return value descriptions:

0	open code
>0	nesting level

See the following example and explanations that follow it.

```
8          %macro A;
9             %put %sysmexecdepth;
10         %mend A;    /* The macro execution depth
                                        of a macro called from open code */
11         %A;          /* is one  */
1
12
13         %macro B;
14            %put %nrstr(%%)sysmexecdepth=%sysmexecdepth;
15            %put %nrstr(%%)sysmexecname(1)=%sysmexecname(1);
16            %put %nrstr(%%)sysmexecname(2)=%sysmexecname(2);
17            %put %nrstr(%%)sysmexecname(0)=%sysmexecname(0);
18            %put %nrstr(%%)sysmexecname(%nrstr(%%)sysmexecdepth-1)=
                  %sysmexecname(%sysmexecdepth-1);
19         %mend B;
20
21         %macro C;
22         %B;
23         %mend;
24         %C;
%sysmexecdepth=2
%sysmexecname(1)=C
```

```
%sysmexecname(2)=B
%sysmexecname(0)=OPEN CODE
%sysmexecname(%sysmexecdepth-1)=C
25
26          %macro level1;
27          %level2;
28          %mend;
29          %macro level2;
30          %level3;
31          %mend;
32          %macro level3;
33          %level4;
34          %mend;
35          %macro level4;
36          %do i = %sysmexecdepth+1 %to -1 %by -1;
37             %put %nrstr(%%)sysmexecname(&i)=%sysmexecname(&i);
38          %end;
39          %mend;
40
41          %level1;
WARNING: Argument 1 to %SYSMEXECNAME function is out of range.
%sysmexecname(5)=
%sysmexecname(4)=LEVEL4
%sysmexecname(3)=LEVEL3
%sysmexecname(2)=LEVEL2
%sysmexecname(1)=LEVEL1
%sysmexecname(0)=OPEN CODE
WARNING: Argument 1 to %SYSMEXECNAME function is out of range.
%sysmexecname(-1)=
42
```

- Macro **A** calls macro **B**. Macro **C** calls macro **B**. A call to %SYSMEXECDEPTH placed in macro **C** would return the value **2** for macro **B**.
- If the macro **C** wanted to know the name of the macro that had called it, it could call %SYSMEXECNAME with **%SYSMEXECNAME(%SYSMEXECDEPTH-1** (the value of the *n* argument being %SYSMEXECDEPTH, its own nesting level, minus one). That call to %SYSMEXECNAME would return the value B.

%SYSMEXECNAME Function

Returns the name of the macro executing at a requested nesting level.

Type: Macro Function

Tip: %SYSMEXECNAME and %SYSMEXECDEPTH were implemented to be used together, but it is not required.

See: %SYSMEXECDEPTH function

Syntax

%SYSMEXECNAME (*n*)

Required Argument

n

The nesting level at which you are requesting the macro name.

0	open code
>0	nesting level

Details

The %SYSMEXECNAME function returns the name of the macro executing at the *n* nesting level. The following three scenarios are shown in the example below.

- If *n = 0*, **`open code`** is returned.
- If *n >%SYSMEXECDEPTH*, a null string is returned and a WARNING diagnostic message is issued to the SAS log.
- If *n<0*, a null string is returned and a WARNING diagnostic message is issued to the SAS log.

```
3          %put %sysmexecdepth; /* The macro execution depth of
                                   Open Code is zero */
0
4          %put %sysmexecname(%sysmexecdepth);
OPEN CODE
5          %put %sysmexecname(%sysmexecdepth + 1);
WARNING: Argument 1 to %SYSMEXECNAME function is out of range.

6          %put %sysmexecname(%sysmexecdepth - 1);
WARNING: Argument 1 to %SYSMEXECNAME function is out of range.
```

%SYSFUNC and %QSYSFUNC Functions

Execute SAS functions or user-written functions.

Type: Macro function

Tip: %SYSFUNC and %QSYSFUNC support SAS function names up to 32 characters.

Syntax

%SYSFUNC (*function*(*argument-1* <*...argument-n*>)<, *format*>)

%QSYSFUNC (*function*(*argument-1* <*...argument-n*>)<, *format*>)

Required Arguments

function

is the name of the function to execute. This function can be a SAS function, a function written with SAS/TOOLKIT software, or a function created using the Chapter 18, “FCMP Procedure” in *Base SAS Procedures Guide*. The function cannot be a macro function.

All SAS functions, except those listed in Table 17.2 on page 269, can be used with %SYSFUNC and %QSYSFUNC.

You cannot nest functions to be used with a single %SYSFUNC. However, you can nest %SYSFUNC calls:

```
%let x=%sysfunc(trim(%sysfunc(left(&num))));
```

Syntax for Selected Functions Used with the %SYSFUNC Function on page 367 shows the syntax of SAS functions used with %SYSFUNC that were introduced with SAS 6.12.

argument-1 <...argument-n>
is one or more arguments used by *function*. An argument can be a macro variable reference or a text expression that produces arguments for a function. If *argument* might contain a special character or mnemonic operator, listed below, use %QSYSFUNC.

format
is an optional format to apply to the result of *function*. This format can be provided by SAS, generated by PROC FORMAT, or created with SAS/TOOLKIT. There is no default value for *format*. If you do not specify a *format*, the SAS macro facility does not perform a *format* operation on the result and uses the default of the *function*.

Details

Because %SYSFUNC is a macro function, you do not need to enclose character values in quotation marks as you do in DATA step functions. For example, the arguments to the OPEN function are enclosed in quotation marks when the function is used alone, but do not require quotation marks when used within %SYSFUNC. These statements show the difference:

- `dsid=open("sasuser.houses","i");`
- `dsid=open("&mydata","&mode");`
- `%let dsid = %sysfunc(open(sasuser.houses,i));`
- `%let dsid=%sysfunc(open(&mydata,&mode));`

All arguments in DATA step functions within %SYSFUNC must be separated by commas. You cannot use argument lists preceded by the word OF.

Note: The arguments to %SYSFUNC are evaluated according to the rules of the SAS macro language. This includes both the function name and the argument list to the function. In particular, an empty argument position will not generate a NULL argument, but a zero length argument.

%SYSFUNC does not mask special characters or mnemonic operators in its result. %QSYSFUNC masks the following special characters and mnemonic operators in its result:

```
& % ' " ( ) + - * / < > = ¬ ^ ~ ; , #  blank
AND OR NOT EQ NE LE LT GE GT IN
```

When a function called by %SYSFUNC or %QSYSFUNC requires a numeric argument, the macro facility converts the argument to a numeric value. %SYSFUNC and %QSYSFUNC can return a floating point number when the function that they execute supports floating point numbers.

Table 17.2 *SAS Functions Not Available with %SYSFUNC and %QSYSFUNC*

All Variable Information Functions	ALLCOMB	ALLPERM

DIF	DIM	HBOUND
IORCMSG	INPUT	LAG
LBOUND	LEXCOMB	LEXCOMBI
LEXPERK	LEXPERM	MISSING
PUT	RESOLVE	SYMGET

Note: Instead of INPUT and PUT, which are not available with %SYSFUNC and %QSYSFUNC, use INPUTN, INPUTC, PUTN, and PUTC.

Note: The Variable Information functions include functions such as VNAME and VLABEL. For a complete list, see "Definitions of Functions and CALL Routines" in *SAS Functions and CALL Routines: Reference*.

CAUTION:

Values returned by SAS functions might be truncated. Although values returned by macro functions are not limited to the length imposed by the DATA step, values returned by SAS functions do have that limitation.

Comparisons

%QSYSFUNC masks the same characters as the %NRBQUOTE function.

Examples

Example 1: Formatting the Current Date in a TITLE Statement

This example formats a TITLE statement containing the current date using the DATE function and the WORDDATE. format:

```
title "%sysfunc(date(),worddate.) Absence Report";
```

When the program is executed on July 18, 2008, the statement produces the following TITLE statement:

```
title "July 18, 2008 Absence Report"
```

Example 2: Formatting a Value Produced by %SYSFUNC

In this example, the TRY macro transforms the value of PARM using the PUTN function and the CATEGORY. format.

```
proc format;
  value category
  Low-<0   = 'Less Than Zero'
  0        = 'Equal To Zero'
  0<-high  = 'Greater Than Zero'
  other    = 'Missing';
run;
%macro try(parm);
  %put &parm is %sysfunc(putn(&parm,category.));
%mend;
```

```
%try(1.02)
%try(.)
%try(-.38)
```

When these statements are executed, these lines are written to the SAS log:

```
1.02 is Greater Than Zero
. is Missing
-.38 is Less Than Zero
```

Example 3: Translating Characters

%SYSFUNC executes the TRANSLATE function to translate the Ns in a string to Ps.

```
%let string1 = V01N01-V01N10;
%let string1 = %sysfunc(translate(&string1,P, N));
%put With N translated to P, V01N01-V01N10 is &string1;
```

When these statements are executed, these lines are written to the SAS log:

```
With N translated to P, V01N01-V01N10 is V01P01-V01P10
```

Example 4: Confirming the Existence of a SAS Data Set

The macro CHECKDS uses %SYSFUNC to execute the EXIST function, which checks the existence of a data set:

```
%macro checkds(dsn);
   %if %sysfunc(exist(&dsn)) %then
      %do;
         proc print data=&dsn;
         run;
      %end;
      %else
         %put The data set &dsn does not exist.;
%mend checkds;
%checkds(sasuser.houses)
```

When the program is executed, the following statements will be produced:

```
PROC PRINT DATA=SASUSER.HOUSES;
RUN;
```

Example 5: Determining the Number of Variables and Observations in a Data Set

Many solutions have been generated in the past to obtain the number of variables and observations present in a SAS data set. Most past solutions have used a combination of _NULL_ DATA steps, SET statement with NOBS=, and arrays to obtain this information. Now, you can use the OPEN and ATTRN functions to obtain this information quickly and without interfering with step boundary conditions.

```
%macro obsnvars(ds);
   %global dset nvars nobs;
   %let dset=&ds;
   %let dsid = %sysfunc(open(&dset));
   %if &dsid %then
      %do;
         %let nobs =%sysfunc(attrn(&dsid,NOBS));
         %let nvars=%sysfunc(attrn(&dsid,NVARS));
         %let rc = %sysfunc(close(&dsid));
         %put &dset has &nvars  variable(s) and &nobs observation(s).;
```

```
      %end;
   %else
      %put Open for data set &dset failed - %sysfunc(sysmsg());
%mend obsnvars;
%obsnvars(sasuser.houses)
```

When the program is executed, the following message will appear in the SAS log:

```
sasuser.houses has 6 variable(s) and 15 observation(s).
```

%SYSGET Function

Returns the value of the specified operating environment variable.

Type: Macro function

Syntax

%SYSGET(*environment-variable*)

Required Argument

environment-variable
is the name of an environment variable. The case of *environment-variable* must agree with the case that is stored on the operating environment.

Details

The %SYSGET function returns the value as a character string. If the value is truncated or the variable is not defined on the operating environment, %SYSGET displays a warning message in the SAS log.

You can use the value returned by %SYSGET as a condition for determining further action to take or parts of a SAS program to execute. For example, your program can restrict certain processing or issue commands that are specific to a user.

For details, see the SAS documentation for your operating environment.

Example: Using SYSGET in a UNIX Operating Environment

This example returns the ID of a user on a UNIX operating environment:

```
%let person=%sysget(USER);
%put User is &person;
```

When these statements execute for user ABCDEF, the following is written to the SAS log:

```
User is abcdef
```

%SYSPROD Function

Reports whether a SAS software product is licensed at the site.

Type: Macro function

See: "%SYSEXEC Statement" on page 316, "SYSSCP and SYSSCPL Automatic Macro Variables" on page 212, and "SYSVER Automatic Macro Variable" on page 218

Syntax

%SYSPROD (*product*)

Required Argument

product

can be a character string or text expression that yields a code for a SAS product. The following are commonly used codes:

Table 17.3 Commonly Used Codes

AF	CPE	GRAPH	PH-CLINICAL
ASSIST	EIS	IML	QC
BASE	ETS	INSIGHT	SHARE
CALC	FSP	LAB	STAT
CONNECT	GIS	OR	TOOLKIT

For codes for other SAS software products, see your on-site SAS support personnel.

Details

%SYSPROD can return the following values:

Table 17.4 %SYSPROD Values and Descriptions

Value	Description
1	The SAS product is licensed.
0	The SAS product is not licensed.
−1	The product is not Institute software (for example, if the product code is misspelled).

Example: Verifying SAS/GRAPH Installation Before Running the GPLOT Procedure

This example uses %SYSPROD to determine whether to execute a PROC GPLOT statement or a PROC PLOT statement, based on whether SAS/GRAPH software has been installed.

```
%macro runplot(ds);
   %if %sysprod(graph)=1 %then
      %do;
         title "GPLOT of %upcase(&ds)";
         proc gplot data=&ds;
            plot style*price / haxis=0 to 150000 by 50000;
         run;
         quit;
      %end;
   %else
      %do;
         title "PLOT of %upcase(&ds)";
         proc plot data=&ds;
            plot style*price;
         run;
         quit;
      %end;
%mend runplot;
%runplot(sasuser.houses)
```

When this program executes and SAS/GRAPH is installed, the following statements are generated:

```
TITLE "GPLOT of SASUSER.HOUSES";
PROC GPLOT DATA=SASUSER.HOUSES;
PLOT STYLE*PRICE / HAXIS=0 TO 150000 BY 50000;
RUN;
```

%UNQUOTE Function

During macro execution, unmasks all special characters and mnemonic operators for a value.

Type: Macro function

See: "%BQUOTE and %NRBQUOTE Functions" on page 242, "%NRBQUOTE Function" on page 246, "%NRQUOTE Function" on page 247, "%NRSTR Function" on page 247, "%QUOTE and %NRQUOTE Functions" on page 248, "%STR and %NRSTR Functions" on page 254, and "%SUPERQ Function" on page 259

Syntax

%UNQUOTE (*character string* | *text expression*)

Details

The %UNQUOTE function unmasks a value so that special characters that it might contain are interpreted as macro language elements instead of as text. The most important effect of %UNQUOTE is to restore normal tokenization of a value whose tokenization was altered by a previous macro quoting function. %UNQUOTE takes effect during macro execution.

For more information, see "Macro Quoting" on page 80.

Example: Using %UNQUOTE to Unmask Values

This example demonstrates a problem that can arise when the value of a macro variable is assigned using a macro quoting function and then the variable is referenced in a later DATA step. If the value is not unmasked before it reaches the SAS compiler, the DATA step does not compile correctly and it produces error messages. Although several macro functions automatically unmask values, a variable might not be processed by one of those functions.

The following program generates error messages in the SAS log because the value of TESTVAL is still masked when it reaches the SAS compiler.

```
%let val = aaa;
%let testval = %str(%'&val%');
data _null_;
  val = &testval;
  put 'VAL =' val;
run;
```

This version of the program runs correctly because %UNQUOTE unmasks the value of TESTVAL.

```
%let val = aaa;
%let testval = %str(%'&val%');
data _null_;
  val = %unquote(&testval);
  put 'VAL =' val;
run;
```

This program prints the following to the SAS log:

```
VAL=aaa
```

%UPCASE and %QUPCASE Functions

Convert values to uppercase.

Type: Macro function

See: "%LOWCASE and %QLOWCASE Autocall Macros" on page 180, "%NRBQUOTE Function" on page 246, and "%QLOWCASE Autocall Macro" on page 182

Syntax

%UPCASE (*character string* | *text expression*)

%QUPCASE(*character string* | *text expression*)

Details

The %UPCASE and %QUPCASE functions convert lowercase characters in the argument to uppercase. %UPCASE does not mask special characters or mnemonic operators in its result, even when the argument was previously masked by a macro quoting function.

If the argument contains a special character or mnemonic operator, listed below, use %QUPCASE. %QUPCASE masks the following special characters and mnemonic operators in its result:

```
& % ' " ( ) + - * / < > = ¬ ^ ~ ; , # blank
AND OR NOT EQ NE LE LT GE GT IN
```

%UPCASE and %QUPCASE are useful in the comparison of values because the macro facility does not automatically convert lowercase characters to uppercase before comparing values.

Comparisons

- %QUPCASE masks the same characters as the %NRBQUOTE function.
- To convert characters to lowercase, use the %LOWCASE or %QLOWCASE autocall macro.

Examples

Example 1: Capitalizing a Value to Be Compared

In this example, the macro RUNREPT compares a value input for the macro variable MONTH to the string DEC. If the uppercase value of the response is DEC, then PROC FSVIEW runs on the data set REPORTS.ENDYEAR. Otherwise, PROC FSVIEW runs on the data set with the name of the month in the REPORTS data library.

```
%macro runrept(month);
   %if %upcase(&month)=DEC %then
       %str(proc fsview data=reports.endyear; run;);
   %else %str(proc fsview data=reports.&month; run;);
%mend runrept;
```

You can invoke the macro in any of these ways to satisfy the %IF condition:

```
%runrept(DEC)
%runrept(Dec)
%runrept(dec)
```

Example 2: Comparing %UPCASE and %QUPCASE

These statements show the results produced by %UPCASE and %QUPCASE:

```
%let a=begin;
%let b=%nrstr(&a);
%put UPCASE produces: %upcase(&b);
%put QUPCASE produces: %qupcase(&b);
```

When these statements execute, the following is written to the SAS log:

```
UPCASE produces: begin
QUPCASE produces: &A
```

Chapter 18

SQL Clauses for Macros

SQL Clauses for Macros

Structured Query Language (SQL) is a standardized, widely used language for retrieving and updating data in databases and relational tables.

Dictionary

INTO Clause

Assigns values produced by PROC SQL to macro variables.

Type: SELECT statement, PROC SQL

Syntax

INTO : *macro-variable-specification-1* < ..., : *macro-variable-specification-n*>

Required Argument

macro-variable-specification

names one or more macro variables to create or update. Precede each macro variable name with a colon (:). The macro variable specification can be in any one or more of the following forms:

: *macro-variable*

specify one or more macro variables. Leading and trailing blanks are not trimmed from values before they are stored in macro variables:

```
select style, sqfeet
   into :type, :size
   from sasuser.houses;
```

:macro-variable-1 – : *macro-variable-n* <NOTRIM>
:macro-variable-1 THROUGH : *macro-variable-n* <NOTRIM>
:macro-variable-1 THRU : *macro-variable-n* <NOTRIM>

specifies a numbered list of macro variables. Leading and trailing blanks are trimmed from values before they are stored in macro variables. If you do not want the blanks to be trimmed, use the NOTRIM option. NOTRIM is an option in each individual element in this form of the INTO clause, so you can use it on one element and not on another element:

```
select style, sqfeet
   into :type1 - :type4 notrim, :size1 - :size3
   from sasuser.houses;
```

:macro-variable SEPARATED BY '*characters* ' <NOTRIM>

specifies one macro variable to contain all the values of a column. Values in the list are separated by one or more *characters*. This form of the INTO clause is useful for building a list of items. Leading and trailing blanks are trimmed from values before they are stored in the macro variable. If you do not want the blanks to be trimmed, use the NOTRIM option. You can use the DISTINCT option in the SELECT statement to store only the unique column (variable) values:

```
select distinct style
   into :types separated by ','
   from sasuser.houses;
```

Details

The INTO clause for the SELECT statement can assign the result of a calculation or the value of a data column (variable) to a macro variable. If the macro variable does not exist, INTO creates it. You can check the PROC SQL macro variable SQLOBS to see the number of rows (observations) produced by a SELECT statement.

The INTO clause can be used only in the outer query of a SELECT statement and not in a subquery. The INTO clause cannot be used when you are creating a table (CREATE TABLE) or a view (CREATE VIEW).

Macro variables created with INTO follow the scoping rules for the %LET statement. For more information, see .

Values assigned by the INTO clause use the BEST12. format.

Comparisons

In the SQL procedure, the INTO clause performs a role similar to the SYMPUT routine.

Examples

Example 1: Storing Column Values in Declared Macro Variables

This example is based on the data set SASUSER.HOUSES and stores the values of columns (variables) STYLE and SQFEET from the first row of the table (or observation in the data set) in macro variables TYPE and SIZE. The %LET statements strip trailing blanks from TYPE and leading blanks from SIZE because this type of specification with INTO does not strip those blanks by default.

```
proc sql noprint;
   select style, sqfeet
      into :type, :size
      from sasuser.houses;
```

```
%let type=&type;
%let size=&size;
%put The first row contains a &type with &size square feet.;
```

When this program executes, the following is written to the SAS log:

```
The first row contains a RANCH with 1250 square feet.
```

Example 2: Storing Row Values in a List of Macro Variables

This example creates two lists of macro variables, TYPE1 through TYPE4 and SIZE1 through SIZE4, and stores values from the first four rows (observations) of the SASUSER.HOUSES data set in them. The NOTRIM option for TYPE1 through TYPE4 retains the trailing blanks for those values.

```
proc sql noprint;
   select style, sqfeet
      into :type1 - :type4 notrim, :size1 - :size4
      from sasuser.houses;
%macro putit;
   %do i=1 %to 4;
      %put Row&i: Type=**&&type&i**   Size=**&&size&i**;
   %end;
%mend putit;
%putit
```

When this program executes, these lines are written to the SAS log:

```
Row1: Type=**RANCH    **  Size=**1250**
Row2: Type=**SPLIT    **  Size=**1190**
Row3: Type=**CONDO    **  Size=**1400**
Row4: Type=**TWOSTORY**  Size=**1810**
```

Example 3: Storing Values of All Rows in one Macro Variable

This example stores all values of the column (variable) STYLE in the macro variable TYPES and separates the values with a comma and a blank.

```
proc sql;
   select distinct quote(style)
      into :types separated by ', '
      from sasuser.houses;
%put Types of houses=&types.;
```

When this program executes, this line is written to the SAS log:

```
Types of houses=CONDO, RANCH, SPLIT, TWOSTORY
```

Chapter 19
Macro Statements

Macro Statements

A macro language statement instructs the macro processor to perform an operation. It consists of a string of keywords, SAS names, and special characters and operators, and it ends in a semicolon. Some macro language statements are used only in macro definitions, but you can use others anywhere in a SAS session or job, either inside or outside macro definitions (referred to as open code).

Dictionary

%ABORT Statement

Stops the macro that is executing along with the current DATA step, SAS job, or SAS session.

Type: Macro statement

Restriction: Allowed in macro definitions only

Syntax

%ABORT <ABEND | CANCEL <FILE> | RETURN | <*n*>> ;

Required Arguments

ABEND

causes abnormal termination of the current macro and SAS job or session. Results depend on the method of operation:

- batch mode and noninteractive mode.
  - stops processing immediately.
  - sends an error message to the SAS log that states that execution was terminated by the ABEND option of the %ABORT macro statement
  - does not execute any subsequent statements or check syntax
  - returns control to the operating environment. Further action is based on how your operating environment and your site treat jobs that end abnormally.
- windowing environment and interactive line mode.
  - causes your macro, windowing environment, and interactive line mode to stop processing immediately and return you to your operating environment.

CANCEL <FILE>

causes the cancellation of the current submitted statements. The results depend on the method of operation.

If the method of operation is batch mode and noninteractive mode, use the CANCEL option to do the following:

- The entire SAS program and SAS system are terminated.
- The error message is written to the SAS log.

If the method of operation is windowing environment and interactive line mode, use the CANCEL option to do the following:

- It only clears the current submitted program.
- Other subsequent submitted programs are not affected.
- The error message is written to the SAS log.

If the method of operation is workspace server and stored process server, use the CANCEL option to do the following:

- It only clears currently submitted program.
- Other subsequent submit calls are not affected.
- The error message is written to the SAS log.

If the method of operation is SAS IntrNet application server, use the CANCEL option to do the following:

- A separate execution is created for each request. The execution submits the request code. A CANCEL in the request code clears the current submitted code but does not terminate the execution or the SAS session.

FILE
when coded as an option to the CANCEL argument in an autoexec file or in a %INCLUDE file, causes only the contents of the autoexec file or %INCLUDE file to be cleared by the %ABORT statement. Other submitted source statements will be executed after the autoexec or %INCLUDE file.

Restriction: The CANCEL argument cannot be submitted using SAS/SHARE,SAS/CONNECT, or SAS/AF.

CAUTION: When %ABORT CANCEL FILE option is executed within a %INCLUDE file, all open macros are closed and execution resumes at the next source line of code.

RETURN
causes abnormal termination of the current macro and SAS job or session. Results depend on the method of operation:

- batch mode and noninteractive mode
  - stops processing immediately
  - sends an error message to the SAS log that states that execution was terminated by the RETURN option of the %ABORT macro statement
  - does not execute any subsequent statements or check syntax
  - returns control to the operating environment with a condition code indicating an error
- windowing environment and interactive line mode
  - causes your macro, windowing environment, and interactive line mode to stop processing immediately and return you to your operating environment

n

an integer value that enables you to specify a condition code:

- When used with the CANCEL argument, the value is placed in the SYSINFO automatic macro variable.
- When it is NOT used with the CANCEL statement, SAS returns the value to the operating environment when the execution stops. The range of values for *n* depends on your operating environment.

Details

If you specify no argument, the %ABORT macro statement produces these results under the following methods of operation:

- batch mode and noninteractive mode.
  - stops processing the current macro and DATA step and writes an error message to the SAS log. Data sets can contain an incomplete number of observations or no observations, depending on when SAS encountered the %ABORT macro statement.
  - sets the OBS= system option to 0.
  - continues limited processing of the remainder of the SAS job, including executing macro statements, executing system option statements, and syntax checking of program statements.
- windowing environment
  - stops processing the current macro and DATA step
  - creates a data set that contains the observations that are processed before the %ABORT macro statement is encountered
  - prints a message to the log that an %ABORT macro statement terminated the DATA step
- interactive line mode
  - stops processing the current macro and DATA step. Any further DATA steps or procedures execute normally.

Comparisons

The %ABORT macro statement causes SAS to stop processing the current macro and DATA step. What happens next depends on

- the method that you use to submit your SAS statements
- the arguments that you use with %ABORT
- your operating environment

The %ABORT macro statement usually appears in a clause of an %IF-%THEN macro statement that is designed to stop processing when an error condition occurs.

Note: The return code generated by the %ABORT macro statement is ignored by SAS if the system option ERRORABEND is in effect.

Note: When you execute an %ABORT macro statement in a DATA step, SAS does not use data sets that were created in the step to replace existing data sets with the same name.

%* Macro Comment Statement

Designates comment text.

Type: Macro statement

Restriction: Allowed in macro definitions or open code

Syntax

%**commentary*;

Required Argument

commentary
is a descriptive message of any length.

Details

The macro comment statement is useful for describing macro code. Text from a macro comment statement is not constant text and is not stored in a compiled macro. Because a semicolon ends the comment statement, the comment cannot contain internal semicolons unless the internal semicolons are enclosed in quotation marks. Macro comment statements are not recognized when they are enclosed in quotation marks.

Macro comment statements are complete macro statements and are processed by the macro facility. Quotation marks within a macro comment must match.

Only macro comment statements and SAS comments of the form /**commentary**/ in macro definitions or open code might be used to hide macro statements from processing by the macro facility.

Comparisons

SAS comment statements of the form

```
*commentary;
```

or

```
comment commentary;
```

are complete SAS statements. Consequently, they are processed by the tokenizer and macro facility and cannot contain semicolons or unmatched quotation marks. SAS comment statements of the form

```
*commentary;
```

or

```
comment commentary;
```

are stored as constant text in a compiled macro. These two types will execute any macro statements within a comment. SAS recommends not to use these within a macro definition.

SAS comments in the form

```
/*commentary*/
```

are not tokenized, but are processed as a string of characters. These comments can appear anywhere a single blank can appear and can contain semicolons or unmatched quotation marks. SAS comments in the form

```
/*commentary*/
```

are not stored in a compiled macro.

Example: Contrasting Comment Types

This code defines and invokes the macro VERDATA, which checks for data errors. It contains a macro comment statement and SAS comments in the form **/*commentary*/** and ***commentary;**

```
%macro verdata(in, thresh);
    *%let thresh = 5;
    /* The preceding SAS comment does not hide the %let statement
        as does this type of SAS comment.
        %let thresh = 6;
    */
   %if %length(&in) > 0 %then %do;
         %* infile given;
      data check;
           /* Jim's data */
         infile &in;
         input x y z;
            * check data;
         if x<&thresh or y<&thresh or z<&thresh then list;
      run;
   %end;
   %else %put Error: No infile specified;
%mend verdata;
%verdata(ina, 0)
```

When you execute VERDATA, the macro processor generates the following:

```
DATA CHECK;
   INFILE INA;
   INPUT X Y Z;
      * CHECK DATA;
   IF X<5 OR Y<5 OR Z<5 THEN LIST;
RUN;
```

%COPY Statement

Copies specified items from a SAS macro library.

Type: Macro statement

Restriction: Allowed in macro definitions or open code

See: "%MACRO Statement" on page 304 and "SASMSTORE= System Option" on page 357

Syntax

%COPY *macro-name* /*<option1 <...option-n>>* SOURCE

Required Arguments

macro-name
name of the macro that the %COPY statement will use.

SOURCESRC

specifies that the source code of the macro will be copied to the output destination. If the OUTFILE= option is not specified, the source is written to the SAS log.

option1 <...option-n>

must be one or more of the following options:

LIBRARY= *libref*LIB=

specifies the libref of a SAS library that contains a catalog of stored compiled SAS macros. If no library is specified, the libref specified by the SASMSTORE= option is used.

Restriction: This libref cannot be WORK.

OUTFILE=*fileref* | *'external file'*OUT=

specifies the output destination of the %COPY statement. The value can be a fileref or an external file.

Example: Using %COPY Statement

In the following example, the %COPY statement writes the stored source code to the SAS log:

```
/* commentary */ %macro foobar(arg) /store source
     des="This macro does not do much";
%put arg = &arg;
* this is commentary!!!;
%* this is macro commentary;
%mend /* commentary; */;       /* Further commentary */
NOTE: The macro FOOBAR completed compilation without errors.
%copy foobar/source;
```

The following results are written to the SAS log:

```
%macro foobar(arg) /store source
des="This macro does not do much";
%put arg = &arg;
* this is commentary!!!;
%* this is macro commentary;
%mend /* commentary; */;
```

%DISPLAY Statement

Displays a macro window.

Type: Macro statement

Restriction: Allowed in macro definitions or open code

See: "%WINDOW Statement" on page 321

Syntax

%DISPLAY *window<.group>* <NOINPUT> <BLANK>
<BELL> <DELETE> ;

Required Arguments

***window* <*.group*>**
names the window and group of fields to be displayed. If the window has more than one group of fields, give the complete *window.group* specification. If a window contains a single unnamed group, specify only *window*.

NOINPUT
specifies that you cannot input values into fields displayed in the window. If you omit the NOINPUT option, you can input values into unprotected fields displayed in the window. Use the NOINPUT option when the %DISPLAY statement is inside a macro definition and you want to merge more than one group of fields into a single display. Using NOINPUT in a particular %DISPLAY statement causes the group displayed to remain visible when later groups are displayed.

BLANK
clears the display in the window. Use the BLANK option to prevent fields from a previous display from appearing in the current display. This option is useful only when the %DISPLAY statement is inside a macro definition and when it is part of a *window.group* specification. When the %DISPLAY statement is outside a macro definition, the display in the window is cleared automatically after the execution of each %DISPLAY statement.

BELL
rings your personal computer's bell, if available, when the window is displayed.

DELETE
deletes the display of the window after processing passes from the %DISPLAY statement on which the option appears. DELETE is useful only when the %DISPLAY statement is inside a macro definition.

Details

You can display only one group of fields in each execution of a %DISPLAY statement. If you display a window containing any unprotected fields, enter values into any required fields and press ENTER to remove the display from the window.

If a window contains only protected fields, pressing ENTER removes the display from the window. While a window is displayed, you can use commands and function keys to view other windows, change the size of the current window, and so on.

%DO Statement

Begins a %DO group.

Type: Macro statement

Restriction: Allowed in macro definitions only

See: "%END Statement" on page 293

Syntax

%DO;
text and macro language statements

%END;

Details

The %DO statement designates the beginning of a section of a macro definition that is treated as a unit until a matching %END statement is encountered. This macro section is called a %DO group. %DO groups can be nested.

A simple %DO statement often appears in conjunction with %IF-%THEN/%ELSE statements to designate a section of the macro to be processed depending on whether the %IF condition is true or false.

Example: Producing One of Two Reports

This macro uses two %DO groups with the %IF-%THEN/%ELSE statement to conditionally print one of two reports.

```
%macro reportit(request);
   %if %upcase(&request)=STAT %then
      %do;
         proc means;
            title "Summary of All Numeric Variables";
         run;
      %end;
   %else %if %upcase(&request)=PRINTIT %then
      %do;
         proc print;
            title "Listing of Data";
         run;
      %end;
   %else %put Incorrect report type. Please try again.;
   title;
%mend reportit;
%reportit(stat)
%reportit(printit)
```

Specifying **`stat`** as a value for the macro variable REQUEST generates the PROC MEANS step. Specifying **`printit`** generates the PROC PRINT step. Specifying any other value writes a customized error message to the SAS log.

%DO, Iterative Statement

Executes a section of a macro repetitively based on the value of an index variable.

Type: Macro statement

Restriction: Allowed in macro definitions only

See: "%END Statement" on page 293

Syntax

%DO *macro-variable=start* %TO *stop* <%BY *increment*> ;
text and macro language statements

%END;

Required Arguments

macro-variable

names a macro variable or a text expression that generates a macro variable name. Its value functions as an index that determines the number of times the %DO loop iterates. If the macro variable specified as the index does not exist, the macro processor creates it in the local symbol table.

You can change the value of the index variable during processing. For example, using conditional processing to set the value of the index variable beyond the *stop* value when a certain condition is met ends processing of the loop.

startstop

specify integers or macro expressions that generate integers to control the number of times the portion of the macro between the iterative %DO and %END statements is processed.

The first time the %DO group iterates, *macro-variable* is equal to *start*. As processing continues, the value of *macro-variable* changes by the value of *increment* until the value of *macro-variable* is outside the range of integers included by *start* and *stop*.

increment

specifies an integer (other than 0) or a macro expression that generates an integer to be added to the value of the index variable in each iteration of the loop. By default, *increment* is 1. *Increment* is evaluated before the first iteration of the loop. Therefore, you cannot change it as the loop iterates.

Example: Generating a Series of DATA Steps

This example illustrates using an iterative %DO group in a macro definition.

```
%macro create(howmany);
   %do i=1 %to &howmany;
      data month&i;
         infile in&i;
         input product cost date;
      run;
   %end;
%mend create;
%create(3)
```

When you execute the macro CREATE, it generates these statements:

```
DATA MONTH1;
   INFILE IN1;
   INPUT PRODUCT COST DATE;
RUN;
DATA MONTH2;
   INFILE IN2;
   INPUT PRODUCT COST DATE;
RUN;
DATA MONTH3;
   INFILE IN3;
   INPUT PRODUCT COST DATE;
RUN;
```

%DO %UNTIL Statement

Executes a section of a macro repetitively until a condition is true.

Type: Macro statement

Restriction: Allowed in macro definitions only

See: "%END Statement" on page 293

Syntax

%DO %UNTIL (*expression*);
text and macro language statements

%END;

Required Argument

expression

can be any macro expression that resolves to a logical value. The macro processor evaluates the expression at the bottom of each iteration. The expression is true if it is an integer other than zero. The expression is false if it has a value of zero. If the expression resolves to a null value or a value containing nonnumeric characters, the macro processor issues an error message.

These examples illustrate expressions for the %DO %UNTIL statement:

- %do %until(&hold=no);
- %do %until(%index(&source,&excerpt)=0);

Details

The %DO %UNTIL statement checks the value of the condition at the bottom of each iteration. Thus, a %DO %UNTIL loop always iterates at least once.

Example: Validating a Parameter

This example uses the %DO %UNTIL statement to scan an option list to test the validity of the parameter TYPE.

```
%macro grph(type);
   %let type=%upcase(&type);
   %let options=BLOCK HBAR VBAR;
   %let i=0;
   %do %until (&type=%scan(&options,&i) or (&i>3)) ;
      %let i = %eval(&i+1);
   %end;
   %if &i>3 %then %do;
      %put ERROR: &type type not supported;
   %end;
   %else %do;
      proc chart;&type sex / group=dept;
      run;
```

```
    %end;
%mend grph;
```

When you invoke the GRPH macro with a value of HBAR, the macro generates these statements:

```
PROC CHART;
HBAR SEX / GROUP=DEPT;
RUN;
```

When you invoke the GRPH macro with a value of PIE, then the %PUT statement writes this line to the SAS log:

```
ERROR: PIE type not supported
```

%DO %WHILE Statement

Executes a section of a macro repetitively while a condition is true.

Type: Macro statement

Restriction: Allowed in macro definitions only

See: “%END Statement” on page 293

Syntax

%DO %WHILE (*expression*);
text and macro program statements

%END;

Required Argument

expression
can be any macro expression that resolves to a logical value. The macro processor evaluates the expression at the top of each iteration. The expression is true if it is an integer other than zero. The expression is false if it has a value of zero. If the expression resolves to a null value or to a value containing nonnumeric characters, the macro processor issues an error message.

These examples illustrate expressions for the %DO %WHILE statement:

- %do %while(&a<&b);
- %do %while(%length(&name)>20);

Details

The %DO %WHILE statement tests the condition at the top of the loop. If the condition is false the first time the macro processor tests it, the %DO %WHILE loop does not iterate.

Example: Removing Markup Tags from a Title

This example demonstrates using the %DO %WHILE to strip markup (SGML) tags from text to create a TITLE statement:

```
%macro untag(title);
      %let stbk=%str(<);
      %let etbk=%str(>);
      /* Do loop while tags exist  */
   %do %while (%index(&title,&stbk)>0) ;
      %let pretag=;
      %let posttag=;
      %let pos_et=%index(&title,&etbk);
      %let len_ti=%length(&title);
          /* Is < first character? */
      %if (%qsubstr(&title,1,1)=&stbk) %then %do;
         %if (&pos_et ne &len_ti) %then
            %let posttag=%qsubstr(&title,&pos_et+1);
      %end;
      %else %do;
         %let pretag=%qsubstr(&title,1,(%index(&title,&stbk)-1));
            /* More characters beyond end of tag (>) ? */
         %if (&pos_et ne &len_ti) %then
            %let posttag=%qsubstr(&title,&pos_et+1);
      %end;
         /* Build title with text before and after tag */
      %let title=&pretag&posttag;
   %end;
title "&title";
%mend untag;
```

You can invoke the macro UNTAG as

```
%untag(<title>Total <emph>Overdue </emph>Accounts</title>)
```

The macro then generates this TITLE statement:

```
TITLE "Total Overdue Accounts";
```

If the title text contained special characters such as commas, you could invoke it with the %NRSTR function.

```
%untag(
   %nrstr(<title>Accounts: Baltimore, Chicago, and Los Angeles</title>))
```

%END Statement

Ends a %DO group.

Type: Macro statement

Restriction: Allowed in macro definitions only

Syntax

%END;

Example: Ending a %DO Group

This macro definition contains a %DO %WHILE loop that ends, as required, with a %END statement:

```
%macro test(finish);
   %let i=1;
   %do %while (&i<&finish);
      %put the value of i is &i;
      %let i=%eval(&i+1);
   %end;
%mend test;
%test(5)
```

Invoking the TEST macro with 5 as the value of *finish* writes these lines to the SAS log:

```
The value of i is 1
The value of i is 2
The value of i is 3
The value of i is 4
```

%GLOBAL Statement

Creates macro variables that are available during the execution of an entire SAS session.

Type: Macro statement

Restriction: Allowed in macro definitions or open code

See: "%LOCAL Statement" on page 302

Syntax

%GLOBAL *macro-variable-1* <*...macro-variable-n*>;

Required Argument

***macro-variable-1* <*..macro-variable-n*>**
is the name of one or more macro variables or a text expression that generates one or more macro variable names. You cannot use a SAS variable list or a macro expression that generates a SAS variable list in a %GLOBAL statement.

Details

The %GLOBAL statement creates one or more global macro variables and assigns null values to the variables. Global macro variables are variables that are available during the entire execution of the SAS session or job.

A macro variable created with a %GLOBAL statement has a null value until you assign it some other value. If a global macro variable already exists and you specify that variable in a %GLOBAL statement, the existing value remains unchanged.

Comparisons

- Both the %GLOBAL statement and the %LOCAL statement create macro variables with a specific scope. However, the %GLOBAL statement creates global macro variables that exist for the duration of the session or job. The %LOCAL statement creates local macro variables that exist only during the execution of the macro that defines the variable.

- If you define both a global macro variable and a local macro variable with the same name, the macro processor uses the value of the local variable during the execution of the macro that contains the local variable. When the macro that contains the local variable is not executing, the macro processor uses the value of the global variable.

Example: Creating Global Variables in a Macro Definition

```
%macro vars(first=1,last=);
   %global gfirst glast;
   %let gfirst=&first;
   %let glast=&last;
   var test&first-test&last;
%mend vars;
```

When you submit the following program, the macro VARS generates the VAR statement and the values for the macro variables used in the title statement.

```
proc print;
   %vars(last=50)
   title "Analysis of Tests &gfirst-&glast";
run;
```

SAS sees the following:

```
PROC PRINT;
   VAR TEST1-TEST50;
   TITLE "Analysis of Tests 1-50";
RUN;
```

%GOTO Statement

Branches macro processing to the specified label.

Type: Macro statement

Alias: %GO TO

Restriction: Allowed in macro definitions only

See: "%label Statement" on page 300

Syntax

%GOTO *label*;

Required Argument

label
is either the name of the label that you want execution to branch to or a text expression that generates the label. A text expression that generates a label in a %GOTO statement is called a *computed %GOTO destination.*[1]

The following examples illustrate how to use *label*:

- `%goto findit;   /* branch to the label FINDIT */`

1 A computed %GOTO contains % or & and resolves to a label.

- ```
  %goto &home;    /* branch to the label that is */
                  /* the value of the macro variable HOME */
  ```

CAUTION:

No percent sign (%) precedes the label name in the %GOTO statement. The syntax of the %GOTO statement does not include a % in front of the label name. If you use a %, the macro processor attempts to call a macro by that name to generate the label.

Details

Branching with the %GOTO statement has two restrictions. First, the label that is the target of the %GOTO statement must exist in the current macro; you cannot branch to a label in another macro with a %GOTO statement. Second, a %GOTO statement cannot cause execution to branch to a point inside an iterative %DO, %DO %UNTIL, or %DO %WHILE loop that is not currently executing.

Example: Providing Exits in a Large Macro

The %GOTO statement is useful in large macros when you want to provide an exit if an error occurs.

```
%macro check(parm);
   %local status;
   %if &parm= %then %do;
       %put ERROR:  You must supply a parameter to macro CHECK.;
       %goto exit;
   %end;
   more macro statements that test for error conditions
   %if &status > 0 %then %do;
       %put ERROR:  File is empty.;
       %goto exit;
   %end;
   more macro statements that generate text
   %put Check completed successfully.;
%exit: %mend check;
```

%IF-%THEN/%ELSE Statement

Conditionally process a portion of a macro.

Type: Macro statement

Restriction: Allowed in macro definitions only

Syntax

%IF *expression* **%THEN** *action*;

<%ELSE*action*;>
```

Required Arguments

expression

is any macro expression that resolves to an integer. If the expression resolves to an integer other than zero, the expression is true and the %THEN clause is processed. If the expression resolves to zero, then the expression is false and the %ELSE statement, if one is present, is processed. If the expression resolves to a null value or a value containing nonnumeric characters, the macro processor issues an error message. For more information about writing macro expressions and their evaluation, see "Macro Expressions" on page 71.

The following examples illustrate using expressions in the %IF-%THEN statement:

- ```
  %if &name=GEORGE %then %let lastname=smith;
  ```
- ```
  %if %upcase(&name)=GEORGE %then %let lastname=smith;
  ```
- ```
  %if &i=10 and &j>5 %then %put check the index variables;
  ```

action

is either constant text, a text expression, or a macro statement. If *action* contains semicolons (for example, in SAS statements), then the first semicolon after %THEN ends the %THEN clause. Use a %DO group or a quoting function, such as %STR, to prevent semicolons in *action* from ending the %IF-%THEN statement. The following examples show two ways to conditionally generate text that contains semicolons:

- ```
  %if &city ne %then %do;
                keep citypop statepop;
             %end;
          %else %do;
                keep statepop;
             %end;
  ```
- ```
  %if &city ne %then %str(keep citypop statepop;);
             %else %str(keep statepop;);
  ```

Details

The macro language does not contain a subsetting %IF statement. Thus, you cannot use %IF without %THEN.

Expressions that compare character values in the %IF-%THEN statement uses the sort sequence of the host operating system for the comparison. Refer to "The SORT PROCEDURE" in the *Base SAS Procedures Guide* for more information about host sort sequences.

Comparisons

Although they look similar, the %IF-%THEN/%ELSE statement and the IF-THEN/ELSE statement belong to two different languages. In general, %IF-%THEN/%ELSE statement, which is part of the SAS macro language, conditionally generates text. However, the IF-THEN/ELSE statement, which is part of the SAS language, conditionally executes SAS statements during DATA step execution.

The expression that is the condition for the %IF-%THEN/%ELSE statement can contain only operands that are constant text or text expressions that generate text. However, the expression that is the condition for the IF-THEN/ELSE statement can contain only operands that are DATA step variables, character constants, numeric constants, or date and time constants.
  ```

When the %IF-%THEN/%ELSE statement generates text that is part of a DATA step, it is compiled by the DATA step compiler and executed. On the other hand, when the IF-THEN/ELSE statement executes in a DATA step, any text generated by the macro facility has been resolved, tokenized, and compiled. No macro language elements exist in the compiled code. “Example 1: Contrasting the %IF-%THEN/%ELSE Statement with the IF-THEN/ELSE Statement” illustrates this difference.

For more information, see “SAS Programs and Macro Processing” on page 11 and “Macro Expressions” on page 71.

Examples

Example 1: Contrasting the %IF-%THEN/%ELSE Statement with the IF-THEN/ELSE Statement

In the SETTAX macro, the %IF-%THEN/%ELSE statement tests the value of the macro variable TAXRATE to control the generation of one of two DATA steps. The first DATA step contains an IF-THEN/ELSE statement that uses the value of the DATA step variable SALE to set the value of the DATA step variable TAX.

```
%macro settax(taxrate);
   %let taxrate = %upcase(&taxrate);
   %if &taxrate = CHANGE %then
      %do;
         data thisyear;
            set lastyear;
            if  sale > 100 then tax = .05;
            else tax = .08;
         run;
      %end;
   %else %if &taxrate = SAME %then
      %do;
         data thisyear;
            set lastyear;
            tax = .03;
            run;
      %end;
%mend settax;
```

When the value of the macro variable TAXRATE is **CHANGE**, then the macro generates the following DATA step:

```
DATA THISYEAR;
   SET LASTYEAR;
   IF SALE > 100 THEN TAX = .05;
   ELSE TAX = .08;
RUN;
```

When the value of the macro variable TAXRATE is **SAME**, then the macro generates the following DATA step:

```
DATA THISYEAR;
   SET LASTYEAR;
   TAX = .03;
RUN;
```

Example 2: Conditionally Printing Reports

In this example, the %IF-%THEN/%ELSE statement generates statements to produce one of two reports.

```
%macro fiscal(report);
   %if %upcase(&report)=QUARTER %then
      %do;
         title 'Quarterly Revenue Report';
         proc means data=total;
            var revenue;
         run;
      %end;
   %else
      %do;
         title 'To-Date Revenue Report';
         proc means data=current;
            var revenue;
         run;
      %end;
%mend fiscal;
%fiscal(quarter)
```

When invoked, the macro FISCAL generates these statements:

```
TITLE 'Quarterly Revenue Report';
PROC MEANS DATA=TOTAL;
VAR REVENUE;
RUN;
```

%INPUT Statement

Supplies values to macro variables during macro execution.

Type: Macro statement

Restriction: Allowed in macro definitions or open code

See: "%PUT Statement" on page 310"%WINDOW Statement" on page 321 and "SYSBUFFR Automatic Macro Variable" on page 192

Syntax

%INPUT<*macro-variable-1* <*...macro-variable-n*>> ;

Required Arguments

no argument
specifies that all text entered is assigned to the automatic macro variable SYSBUFFR.

***macro-variable-1* <*...macro-variable-n*>**
is the name of a macro variable or a macro text expression that produces a macro variable name. The %INPUT statement can contain any number of variable names separated by blanks.

Details

The macro processor interprets the line submitted immediately after a %INPUT statement as the response to the %INPUT statement. That line can be part of an interactive line mode session, or it can be submitted from within the Program Editor window during a windowing environment session.

When a %INPUT statement executes as part of an interactive line mode session, the macro processor waits for you to enter a line containing values. In a windowing environment session, the macro processor does NOT wait for you to input values. Instead, it simply reads the next line that is processed in the program and attempts to assign variable values. Likewise, if you invoke a macro containing a %INPUT statement in open code as part of a longer program in a windowing environment, the macro processor reads the next line in the program that follows the macro invocation. When you submit a %INPUT statement in open code from a windowing environment, ensure that the line that follows a %INPUT statement or a macro invocation that includes a %INPUT statement contains the values that you want to assign.

When you name variables in the %INPUT statement, the macro processor matches the variables with the values in your response based on their positions. That is, the first value that you enter is assigned to the first variable named in the %INPUT statement, the second value is assigned to the second variable, and so on.

Each value to be assigned to a particular variable must be a single word or a string enclosed in quotation marks. To separate values, use blanks. After all values have been matched with macro variable names, excess text becomes the value of the automatic macro variable SYSBUFFR.

Example: Assigning a Response to a Macro Variable

In an interactive line mode session, the following statements display a prompt and assign the response to the macro variable FIRST:

```
%put Enter your first name:;
%input first;
```

%label Statement

Identifies the destination of a %GOTO statement.

Type: Macro statement

Restriction: Allowed in macro definitions only

See: "%GOTO Statement" on page 295

Syntax

%label: macro-text

Required Arguments

label
specifies a SAS name.

macro-text
is a macro statement, a text expression, or constant text. The following examples illustrate each:

- `%one: %let book=elementary;`
- `%out: %mend;`
- `%final: data _null_;`

Details

- The label name is preceded by a %. When you specify this label in a %GOTO statement, do not precede it with a %.
- An alternative to using the %GOTO statement and statement label is to use a %IF-%THEN statement with a %DO group.

Example: Controlling Program Flow

In the macro INFO, the %GOTO statement causes execution to jump to the label QUICK when the macro is invoked with the value of **`short`** for the parameter TYPE.

```
%macro info(type);
   %if %upcase(&type)=SHORT %then %goto quick; /* No % here */
      proc contents;
      run;
      proc freq;
         tables _numeric_;
      run;
   %quick: proc print data=_last_(obs=10);     /* Use % here */
      run;
%mend info;
%info(short)
```

Invoking the macro INFO with TYPE equal to **`short`** generates these statements:

```
PROC PRINT DATA=_LAST_(OBS=10);
   RUN;
```

%LET Statement

Creates a macro variable and assigns it a value.

Type: Macro statement

Restriction: Allowed in macro definitions or open code

See: "%STR and %NRSTR Functions" on page 254

Syntax

%LET *macro-variable* =<*value*> ;

Required Arguments

macro-variable
is either the name of a macro variable or a text expression that produces a macro variable name. The name can refer to a new or existing macro variable.

value
is a character string or a text expression. Omitting *value* produces a null value (0 characters). Leading and trailing blanks in *value* are ignored. To make them significant, enclose *value* with the %STR function.

Details

If the macro variable named in the %LET statement already exists, the %LET statement changes the value. A %LET statement can define only one macro variable at a time.

Example: Sample %LET Statements

These examples illustrate several %LET statements:

```
%macro title(text,number);
    title&number "&text";
%mend;
%let topic=  The History of Genetics  ; /* Leading and trailing */
                                        /* blanks are removed   */
%title(&topic,1)
%let subject=topic;                     /* &subject resolves    */
%let &subject=Genetics Today;           /* before assignment    */
%title(&topic,2)
%let subject=The Future of Genetics;    /* &subject resolves    */
%let topic= &subject;                   /* before assignment    */
%title(&topic,3)
```

When you submit these statements, the TITLE macro generates the following statements:

```
TITLE1 "The History of Genetics";
TITLE2 "Genetics Today";
TITLE3 "The Future of Genetics";
```

%LOCAL Statement

Creates macro variables that are available only during the execution of the macro where they are defined.

Type: Macro statement

Restriction: Allowed in macro definitions only

See: “%GLOBAL Statement” on page 294

Syntax

%LOCAL *macro-variable-1* <*...macro-variable-n*>;

Required Argument

macro-variable-1 <...macro-variable-n>
is the name of one or more macro variables or a text expression that generates one or more macro variable names. You cannot use a SAS variable list or a macro expression that generates a SAS variable list in a %LOCAL statement.

Details

The %LOCAL statement creates one or more local macro variables. A macro variable created with %LOCAL has a null value until you assign it some other value. Local macro variables are variables that are available only during the execution of the macro in which they are defined.

Use the %LOCAL statement to ensure that macro variables created earlier in a program are not inadvertently changed by values assigned to variables with the same name in the current macro. If a local macro variable already exists and you specify that variable in a %LOCAL statement, the existing value remains unchanged.

Comparisons

- Both the %LOCAL statement and the %GLOBAL statement create macro variables with a specific scope. However, the %LOCAL statement creates local macro variables that exist only during the execution of the macro that contains the variable, and the %GLOBAL statement creates global macro variables that exist for the duration of the session or job.
- If you define a local macro variable and a global macro variable with the same name, the macro facility uses the value of the local variable during the execution of the macro that contains that local variable. When the macro that contains the local variable is not executing, the macro facility uses the value of the global variable.

Example: Using a Local Variable with the Same Name as a Global Variable

```
%let variable=1;
%macro routine;
   %put ***** Beginning ROUTINE *****;
   %local variable;
   %let variable=2;
   %put The value of variable inside ROUTINE is &variable;
   %put ***** Ending ROUTINE *****;
%mend routine;
%routine
%put The value of variable outside ROUTINE is &variable;
```

Submitting these statements writes these lines to the SAS log:

```
***** Beginning ROUTINE *****
The value of variable inside ROUTINE is 2
***** Ending ROUTINE *****
The value of variable outside ROUTINE is 1
```

%MACRO Statement

Begins a macro definition.

Type: Macro statement

Restriction: Allowed in macro definitions or open code

See: "%MEND Statement" on page 310 and "SYSPBUFF Automatic Macro Variable" on page 210

Syntax

%MACRO *macro-name* <*(parameter-list)*> </ *option-1* <*...option-n*>> ;

Required Arguments

macro-name

names the macro. A macro name must be a SAS name, which you supply; you cannot use a text expression to generate a macro name in a %MACRO statement. In addition, do not use macro reserved words as a macro name. (For a list of macro reserved words, see " Reserved Words in the Macro Facility" on page 363.)

parameter-list

names one or more local macro variables whose values you specify when you invoke the macro. Parameters are local to the macro that defines them. You must supply each parameter name; you cannot use a text expression to generate it. A parameter list can contain any number of macro parameters separated by commas. The macro variables in the parameter list are usually referenced in the macro.

- *parameter-list* can be
  - *<positional parameter-1><. . . ,positional parameter-n>*
  - *<keyword-parameter=<value> <. . . ,keyword-parameter-n=<value>>>*

positional-parameter-1 <. . . ,positional-parameter-n>
: specifies one or more positional parameters. You can specify positional parameters in any order, but in the macro invocation, the order in which you specify the values must match the order that you list them in the %MACRO statement. If you define more than one positional parameter, use a comma to separate the parameters. If at invocation that you do not supply a value for a positional parameter, the macro facility assigns a null value to that parameter.

keyword-parameter=<value> <. . . ,keyword-parameter-n=<value>>
: names one or more macro parameters followed by equal signs. You can specify default values after the equal signs. If you omit a default value after an equal sign, the keyword parameter has a null value. Using default values enables you to write more flexible macro definitions and reduces the number of parameters that must be specified to invoke the macro. To override the default value, specify the macro variable name followed by an equal sign and the new value in the macro invocation.

Note: You can define an unlimited number of parameters. If both positional and keyword parameters appear in a macro definition, positional parameters must come first.

option-1 <...option-n>
can be one or more of these optional arguments:

CMD
specifies that the macro can accept either a name-style invocation or a command-style invocation. Macros defined with the CMD option are sometimes called *command-style macros*.

Use the CMD option only for macros that you plan to execute from the command line of a SAS window. The SAS system option CMDMAC must be in effect to use command-style invocations. If CMDMAC is in effect and you have defined a command-style macro in your program, the macro processor scans the first word of every SAS command to see whether it is a command-style macro invocation. When the SAS system option NOCMDMAC option is in effect, the macro processor treats only the words following the % symbols as potential macro invocations. If the CMDMAC option is not in effect, you still can use a name-style invocation for a macro defined with the CMD option.

DES='*text*'
specifies a description for the macro entry in the macro catalog. The description text can be up to 256 characters in length. Enclose the description in quotation marks. This description appears in the CATALOG window when you display the contents of the catalog containing the stored compiled macros. The DES= option is especially useful when you use the stored compiled macro facility.

MINDELIMITER='*single character*';
specifies a value that will override the value of the MINDELIMITER= global option. The value must be a single character enclosed in single quotation marks and can appear only once in a %MACRO statement.

MINOPERATOR | NOMINOPERATOR
specifies that the macro processor recognizes and evaluates the mnemonic `IN` and the special character `#` as logical operators when evaluating arithmetic or logical expressions during the execution of the macro. The setting of this argument overrides the setting of the NOMINOPERATOR global system option.

The NOMINOPERATOR argument specifies that the macro processor does not recognize the mnemonic `IN` and the special character `#` as logical operators when evaluating arithmetic or logical expressions during the execution of the macro. The setting of this argument overrides the setting of the MINOPERATOR global system option.

PARMBUFF
assigns the entire list of parameter values in a macro call, including the parentheses in a name-style invocation, as the value of the automatic macro variable SYSPBUFF. Using the PARMBUFF option, you can define a macro that accepts a varying number of parameter values.

If the macro definition includes both a set of parameters and the PARMBUFF option, the macro invocation causes the parameters to receive values. It also causes the entire invocation list of values to be assigned to SYSPBUFF.

To invoke a macro defined with the PARMBUFF option in a windowing environment or interactive line mode session without supplying a value list, enter an empty set of parentheses or more program statements after the invocation.

This action indicates the absence of a value list, even if the macro definition contains no parameters.

SECURE | NOSECURE

causes the contents of a macro to be encrypted when stored in a stored compiled macro library. This feature enables you to write *secure* macros that will protect intellectual property that is contained in the macros. The macros are secured using the Encryption Algorithm Manager.

A NOSECURE option has been implemented to aid in the global edit of a source file or library to turn on security. For example, when you are creating several macros that will need to be secure. When creating the macros, use the NOSECURE option. When all macros are completed and ready for production, you can do a global edit and change NOSECURE to SECURE.

If you use the SECURE and SOURCE options on a macro, no output is produced when you use the %COPY statement. The following NOTE is written to the SAS log:

```
NOTE: The macro %name was compiled with the SECURE
option. No output will be produced for this %COPY
statement.
```

STMT

specifies that the macro can accept either a name-style invocation or a statement-style invocation. Macros defined with the STMT option are sometimes called *statement-style macros*.

The IMPLMAC system option must be in effect to use statement-style macro invocations. If IMPLMAC is in effect and you have defined a statement-style macro in your program, the macro processor scans the first word of every SAS statement to see whether it is a statement-style macro invocation. When the NOIMPLMAC option is in effect, the macro processor treats only the words following the % symbols as potential macro invocations. If the IMPLMAC option is not in effect, you still can use a name-style invocation for a macro defined with the STMT option.

SOURCE SRC

combines and stores the source of the compiled macro with the compiled macro code as an entry in a SAS catalog in a permanent SAS library. The SOURCE option requires that the STORE option and the MSTORED option be set. You can use the SASMSTORE= option to identify a permanent SAS library. You can store a macro or call a stored compiled macro only when the MSTORED option is in effect. (For more information, see “Storing and Reusing Macros” on page 113.)

Note: The source code saved by the SOURCE option begins with the %MACRO keyword and ends with the semi-colon following the %MEND statement.

CAUTION:

The SOURCE option cannot be used on nested macro definitions (macro definitions contained within another macro).

STORE

stores the compiled macro as an entry in a SAS catalog in a permanent SAS library. Use the SAS system option SASMSTORE= to identify a permanent SAS library. You can store a macro or call a stored compiled macro only when the SAS system option MSTORED is in effect. (For more information, see “Storing and Reusing Macros” on page 113.)

Details

The %MACRO statement begins the definition of a macro, assigns the macro a name, and can include a list of macro parameters, a list of options, or both.

A macro definition must precede the invocation of that macro in your code. The %MACRO statement can appear anywhere in a SAS program, except within data lines. A macro definition cannot contain a CARDS statement, a DATALINES statement, a PARMCARDS statement, or data lines. Use an INFILE statement instead.

By default, a defined macro is an entry in a SAS catalog in the WORK library. You can also store a macro in a permanent SAS catalog for future use. However, in SAS 6 and earlier, SAS does not support copying, renaming, or transporting macros.

You can nest macro definitions, but doing so is rarely necessary and is often inefficient. If you nest a macro definition, then it is compiled every time you invoke the macro that includes it. Instead, nesting a macro invocation inside another macro definition is sufficient in most cases.

Examples

Example 1: Using the %MACRO Statement with Positional Parameters

In this example, the macro PRNT generates a PROC PRINT step. The parameter in the first position is VAR, which represents the SAS variables that appear in the VAR statement. The parameter in the second position is SUM, which represents the SAS variables that appear in the SUM statement.

```
%macro prnt(var,sum);
   proc print data=srhigh;
      var &var;
      sum &sum;
   run;
%mend prnt;
```

In the macro invocation, all text up to the comma is the value of parameter VAR; text following the comma is the value of parameter SUM.

```
%prnt(school district enrollmt, enrollmt)
```

During execution, macro PRNT generates the following statements:

```
PROC PRINT DATA=SRHIGH;
   VAR SCHOOL DISTRICT ENROLLMT;
   SUM ENROLLMT;
RUN;
```

Example 2: Using the %MACRO Statement with Keyword Parameters

In the macro FINANCE, the %MACRO statement defines two keyword parameters, YVAR and XVAR, and uses the PLOT procedure to plot their values. Because the keyword parameters are usually EXPENSES and DIVISION, default values for YVAR and XVAR are supplied in the %MACRO statement.

```
%macro finance(yvar=expenses,xvar=division);
   proc plot data=yearend;
      plot &yvar*&xvar;
```

```
   run;
%mend finance;
```

- To use the default values, invoke the macro with no parameters.

  ```
  %finance()
  ```

  or

  ```
  %finance;
  ```

  The macro processor generates this SAS code:

  ```
  PROC PLOT DATA=YEAREND;
     PLOT EXPENSES*DIVISION;
  RUN;
  ```

- To assign a new value, give the name of the parameter, an equal sign, and the value:

  ```
  %finance(xvar=year)
  ```

  Because the value of YVAR did not change, it retains its default value. Macro execution produces this code:

  ```
  PROC PLOT DATA=YEAREND;
     PLOT EXPENSES*YEAR;
  RUN;
  ```

Example 3: Using the %MACRO Statement with the PARMBUFF Option

The macro PRINTZ uses the PARMBUFF option to enable you to input a different number of arguments each time you invoke it:

```
%macro printz/parmbuff;
   %let num=1;
   %let dsname=%scan(&syspbuff,&num);
   %do %while(&dsname ne);
      proc print data=&dsname;
      run;
      %let num=%eval(&num+1);
      %let dsname=%scan(&syspbuff,&num);
   %end;
%mend printz;
```

This invocation of PRINTZ contains four parameter values, **PURPLE**, **RED**, **BLUE**, and **TEAL** although the macro definition does not contain any individual parameters:

```
%printz(purple,red,blue,teal)
```

As a result, SAS receives these statements:

```
PROC PRINT DATA=PURPLE;
RUN;
PROC PRINT DATA=RED;
RUN;
PROC PRINT DATA=BLUE;
RUN;
PROC PRINT DATA=TEAL;
RUN;
```

Example 4: Using the %MACRO Statement with the SOURCE Option

The SOURCE option combines and stores the source of the compiled macro with the compiled macro code. Use the %COPY statement to write the source to the SAS log. For more information about viewing or retrieving the stored source, see "%COPY Statement" on page 286.

```
/* commentary */  %macro foobar(arg) /store source
      des="This macro does not do much";
%put arg = &arg;
* this is commentary!!!;
%* this is macro commentary;
%mend /* commentary; */;      /* Further commentary */
NOTE: The macro FOOBAR completed compilation without errors.
%copy foobar/source;
```

The following results are written to the SAS log:

```
%macro foobar(arg) /store source
des="This macro does not do much";
%put arg = &arg;
* this is commentary!!!;
%* this is macro commentary;
%mend /* commentary; */;
```

Example 5: Using the %MACRO Statement with the STORE and SECURE Options

The SECURE option can be used only in conjunction with the STORE option. The following example demonstrates the use of the STORE and an implied NOSECURE option to create a macro that is stored in plain text.

```
options mstored sasmstore=mylib;
libname mylib "mylib";
%macro nonsecure/store; /* This macro is stored in plain text */
  data _null_;
    x=1;
    put "This data step was generated from a non-secure macro.";
  run;
%mend nonsecure;
%nonsecure
filename maccat catalog 'mylib.sasmacr.nonsecure.macro';
data _null_;
  infile maccat;
  input;
  list;
run;
```

The following example demonstrates the use of the STORE and SECURE options to create a macro that is encrypted.

```
options mstored sasmstore=mylib;
libname mylib "mylib";
%macro secure/store secure; /* This macro is encrypted */
    data _null_;
      x=1;
      put "This data step was generated from a secure macro.";
    run;
  %mend secure;
```

```
%secure
filename maccat catalog 'mylib.sasmacr.secure.macro';
data _null_;
  infile maccat;
  input;
  list;
run;
```

%MEND Statement

Ends a macro definition.

Type: Macro statement

Restriction: Allowed in macro definitions only

Syntax

%MEND <*macro-name*> ;

Required Argument

macro-name
names the macro as it ends a macro definition. Repeating the name of the macro is optional, but it is useful for clarity. If you specify *macro-name*, the name in the %MEND statement should match the name in the %MACRO statement; otherwise, SAS issues a warning message.

Example: Ending a Macro Definition

```
%macro disc(dsn);
   data &dsn;
      set perm.dataset;
      where month="&dsn";
   run;
%mend disc;
```

%PUT Statement

Writes text or macro variable information to the SAS log.

Type: Macro statement

Restriction: Allowed in macro definitions or open code

Syntax

%PUT <*text* | _ALL_ | _AUTOMATIC_ | _GLOBAL_ | _LOCAL_ | _USER_> ;

Required Arguments

no argument
places a blank line in the SAS log.

text
: is text or a text expression that is written to the SAS log. If *text* is longer than the current line size, the remainder of the text appears on the next line. The %PUT statement removes leading and trailing blanks from *text* unless you use a macro quoting function.

ALL
: lists the values of all user-generated and automatic macro variables.

AUTOMATIC
: lists the values of automatic macro variables. The automatic variables listed depend on the SAS products installed at your site and on your operating system. The scope is identified as AUTOMATIC.

GLOBAL
: lists user-generated global macro variables. The scope is identified as GLOBAL.

LOCAL
: lists user-generated local macro variables. The scope is the name of the currently executing macro.

USER
: describes user-generated global and local macro variables. The scope is identified either as GLOBAL, or as the name of the macro in which the macro variable is defined.

Details

When you use the %PUT statement to list macro variable descriptions, the %PUT statement includes only the macro variables that exist at the time the statement executes. The description contains the macro variable's scope, name, and value. Macro variables with null values show only the scope and name of the variable. Characters in values that have been quoted with macro quoting functions remain quoted. Values that are too long for the current line size wrap to the next line or lines. Macro variables are listed in order from the current local macro variables outward to the global macro variables.

Note: Within a particular scope, macro variables might appear in any order, and the order might change in different executions of the %PUT statement or different SAS sessions. Do not write code that depends on locating a variable in a particular position in the list.

The following figure shows the relationship of these terms.

Figure 19.1 *%PUT Arguments by Type and Scope*

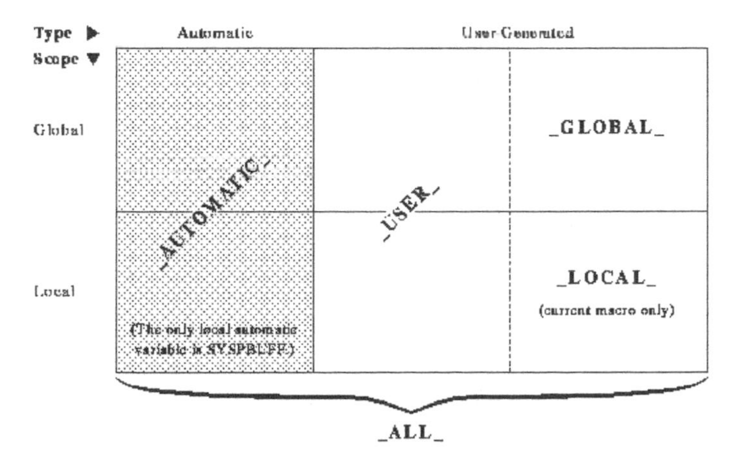

The %PUT statement displays text in different colors to generate messages that look like ERROR, NOTE, and WARNING messages generated by SAS. To display text in different colors, the first word in the %PUT statement must be ERROR, NOTE, or WARNING, followed immediately by a colon or a hyphen. You might also use the national-language equivalents of these words. When you use a hyphen, the ERROR, NOTE, or WARNING word is blanked out.

Note: If you use the %PUT statement and the last message text that was generated by the SYSWARNINGTEXT and SYSERRORTEXT automatic macro variables contained an & or %, you must use the %SUPERQ macro quoting function. For more information, see "SYSERRORTEXT Automatic Macro Variable" on page 202 and "SYSWARNINGTEXT Automatic Macro Variable" on page 219.

TIP If you place an equal sign between the ampersand and the macro variable name of a direct macro variable reference, the macro variable's name displays in the log along with the macro variable's value.

```
%let x=1;
%put &=x;
X=1;
```

Examples

Example 1: Displaying Text

The following statements illustrate using the %PUT statement to write text to the SAS log:

```
%put One line of text.;
%put %str(Use a semicolon(;) to end a SAS statement.);
%put %str(Enter the student%'s address.);
```

When you submit these statements, these lines appear in the SAS log:

```
One line of text.
Use a semicolon(;) to end a SAS statement.
Enter the student's address.
```

Example 2: Displaying Automatic Variables

To display all automatic variables, submit

```
%put _automatic_;
```

The result in the SAS log (depending on the products installed at your site) lists the scope, name, and value of each automatic variable:

```
AUTOMATIC SYSBUFFR
AUTOMATIC SYSCMD
AUTOMATIC SYSDATE 21JUN97
AUTOMATIC SYSDAY Wednesday
AUTOMATIC SYSDEVIC
AUTOMATIC SYSDSN          _NULL_
AUTOMATIC SYSENV FORE
AUTOMATIC SYSERR 0
AUTOMATIC SYSFILRC 0
AUTOMATIC SYSINDEX 0
AUTOMATIC SYSINFO 0
```

Example 3: Displaying User-Generated Variables

This example lists the user-generated macro variables in all scopes.

```
%macro myprint(name);
   proc print data=&name;
      title "Listing of &name on &sysdate";
      footnote "&foot";
   run;
   %put _user_;
%mend myprint;
%let foot=Preliminary Data;
%myprint(consumer)
```

The %PUT statement writes these lines to the SAS log:

```
MYPRINT NAME consumer
GLOBAL FOOT Preliminary Data
```

Notice that SYSDATE does not appear because it is an automatic macro variable.

To display the user-generated variables after macro MYPRINT finishes, submit another %PUT statement.

```
%put _user_;
```

The result in the SAS log does not list the macro variable NAME because it was local to MYPRINT and ceased to exist when MYPRINT finished execution.

```
GLOBAL FOOT Preliminary Data
```

Example 4: Displaying Local Variables

This example displays the macro variables that are local to macro ANALYZE.

```
%macro analyze(name,vars);
   proc freq data=&name;
      tables &vars;
   run;
   %put FIRST LIST:;
   %put _local_;
   %let firstvar=%scan(&vars,1);
   proc print data=&name;
      where &firstvar ne .;
   run;
   %put SECOND LIST:;
   %put _local_;
%mend analyze;
%analyze(consumer,car house stereo)
```

In the result that is printed in the SAS log, the macro variable FIRSTVAR, which was created after the first %PUT _LOCAL_ statement, appears only in the second list.

```
FIRST LIST:
ANALYZE NAME consumer
ANALYZE VARS car house stereo
SECOND LIST:
ANALYZE NAME consumer
ANALYZE VARS car house stereo
ANALYZE FIRSTVAR car
```

%RETURN Statement

Execution causes normal termination of the currently executing macro.

Type: Macro Statement

Restriction: Valid only in a macro definition

Syntax

%RETURN;

Details

The %RETURN macro causes normal termination of the currently executing macro.

Example: Using %RETURN Statement

In this example, if the error variable is set to 1, then the macro will stop executing and the DATA step will not execute.

```
%macro checkit(error);
   %if &error = 1 %then %return;
    data a;
       x=1;
    run;
%mend checkit;
%checkit(0)
%checkit(1)
```

%SYMDEL Statement

Deletes the specified variable or variables from the macro global symbol table.

Type: Macro Statement

Syntax

%SYMDEL *macro-variable-1* <...*macro-variable-n*></option> ;

Required Arguments

***macro-variable-1* <..*macro-variable-n*>**
is the name of one or more macro variables or a text expression that generates one or more macro variable names. You cannot use a SAS variable list or a macro expression that generates a SAS variable list in a %SYMDEL statement.

options

NOWARN
suppresses the warning message when an attempt is made to delete a non-existent macro variable.

Details

%SYMDEL statement issues a warning when an attempt is made to delete a non-existent macro variable. To suppress this message, use the NOWARN option.

%SYSCALL Statement

Invokes a SAS call routine.

Type: Macro statement

Restriction: Allowed in macro definitions or in open code

See: "%SYSFUNC and %QSYSFUNC Functions" on page 268

Syntax

%SYSCALL *call-routine<(call-routine-argument-1 <...call-routine-argument-n>)>* ;

Required Arguments

call-routine

is a SAS or user-written CALL routine created with SAS/TOOLKIT software or a routine created using the "FCMP Procedure" in *Base SAS Procedures Guide*. All SAS call routines are accessible with %SYSCALL except LABEL, VNAME, SYMPUT, and EXECUTE.

call-routine-argument-1 <...call-routine-argument-n>

is one or more macro variable names (with no leading ampersands), separated by commas. You can use a text expression to generate part or all of the CALL routine arguments.

Details

When %SYSCALL invokes a CALL routine, the value of each macro variable argument is retrieved and passed unresolved to the CALL routine. Upon completion of the CALL routine, the value for each argument is written back to the respective macro variable. If %SYSCALL encounters an error condition, the execution of the CALL routine terminates without updating the macro variable values, an error message is written to the log, and macro processing continues.

Note: The arguments to %SYSCALL are evaluated according to the rules of the SAS macro language. This includes both the function name and the argument list to the function. In particular, an empty argument position will not generate a NULL argument, but a zero length argument.

CAUTION:

Do not use leading ampersands on macro variable names. The arguments in the CALL routine invoked by the %SYSCALL macro are resolved before execution. If you use leading ampersands, then the values of the macro variables are passed to the CALL routine rather than the names of the macro variables.

CAUTION:

Macro variables contain only character data. When an argument to a function might be either numeric data or character data, %SYSCALL attempts to convert the supplied data to numeric data. This causes truncation of any trailing blanks if the data was character data. %SYSCALL does not modify arguments that might be character data. You can preserve the trailing blanks by using the %QUOTE function when assigning the value to the macro variable that will be supplied as the argument to the function. To determine whether it is necessary to preserve the trailing blanks using the %QUOTE function, consult the documentation for the desired function to see whether the arguments are numeric only, character only, or either numeric or character. Use the %QUOTE function to quote the value supplied to arguments that are documented to be either numeric or character.

```
%let j=1;
%let x=fax;
%let y=fedex;
%let z=phone;
%put j=&j x=&x y=&y z=&z
j=1 x=fax y=fedex z=phone
%syscall allperm(j,x,y,z);
%put j=&j x=&x y=&y z=&z
j=1 x=250 y=65246 z=phone
```

Example: Using the RANUNI Call Routine with %SYSCALL

This example illustrates the %SYSCALL statement. The macro statement %SYSCALL RANUNI(A,B) invokes the SAS CALL routine RANUNI.

Note: The syntax for RANUNI is **RANUNI(seed,x)**.

```
%let a = 123456;
%let b = 0;
%syscall ranuni(a,b);
%put &a, &b;
```

The %PUT statement writes the following values of the macro variables A and B to the SAS log:

```
1587033266 0.739019954
```

%SYSEXEC Statement

Issues operating environment commands.

Type: Macro statement

Restriction: Allowed in macro definitions or open code

See: "SYSSCP and SYSSCPL Automatic Macro Variables" on page 212 and "SYSRC Automatic Macro Variable" on page 212

Syntax

%SYSEXEC<*command*> ;

Required Arguments

no argument
: puts you into operating environment mode under most operating environments, where you can issue operating environment commands and return to your SAS session.

command
: is any operating environment command. If *command* contains a semicolon, use a macro quoting function.

Details

The %SYSEXEC statement causes the operating environment to immediately execute the command that you specify and assigns any return code from the operating environment to the automatic macro variable SYSRC. Use the %SYSEXEC statement and the automatic macro variables SYSSCP and SYSSCPL to write portable macros that run under multiple operating environments.

Operating Environment Information
: The following items related to the use of the %SYSEXEC statement are operating environment specific. For details, see the SAS documentation for your operating environment.

- the availability of the %SYSEXEC statement in batch processing, noninteractive mode, or interactive line mode.
- the way you return from operating environment mode to your SAS session after executing the %SYSEXEC statement with no argument.
- the commands to use with the %SYSEXEC statement.
- the return codes that you get in the automatic macro variable SYSRC.

Comparisons

The %SYSEXEC statement is analogous to the X statement and the X windowing environment command. However, unlike the X statement and the X windowing environment command, host commands invoked with %SYSEXEC should not be enclosed in quotation marks.

%SYSLPUT Statement

Creates a new macro variable or modifies the value of an existing macro variable on a remote host or server.

Type: Macro Statement

Restriction: Allowed in macro definitions or open code

Requirement: SAS/CONNECT

See: "%LET Statement" on page 301 and "%SYSRPUT Statement" on page 319

Syntax

%SYSLPUT*macro-variable*=<*value*</REMOTE=*remote-session-identifier*>> ;

Required Arguments

macro-variable
is either the name of a macro variable or a macro expression that produces a macro variable name. The name can refer to a new or existing macro variable on a remote host or server.

remote-session-identifier
is the name of the remote session.

value
is a string or a macro expression that yields a string. Omitting the value produces a null (0 characters). Leading and trailing blanks are ignored. To make them significant, enclose the value in the %STR function.

Details

The %SYSLPUT statement is submitted with SAS/CONNECT software from the local host or client to a remote host or server to create a new macro variable on the remote host or server, or to modify the value of an existing macro variable on the remote host or server.

Note: The names of the macro variables on the remote and local hosts must not contain any leading ampersands.

To assign the value of a macro variable on a remote host to a macro variable on the local host, use the %SYSRPUT statement.

To use %SYSLPUT, you must have initiated a link between a local SAS session or client and a remote SAS session or server using the SIGNON command or SIGNON statement. For more information, see the documentation for SAS/CONNECT software.

%SYSMACDELETE Statement

Deletes a macro definition from the WORK.SASMACR catalog.

Type: Macro Statement

Restriction: Allowed in macro definition and open code

Syntax

%SYSMACDELETE *macro_name</ option>*;

Required Argument

macro_name
the name of a macro or a text expression that produces a macro variable name.

Optional Argument

NOWARN
specifies that no warning diagnostic message should be issued.

Details

The %SYSMACDELETE statement deletes the macro definition of the specified macro from the WORK.SASMACR catalog. If no definition for the macro exists in the WORK.SASMACR catalog, a WARNING diagnostic message is issued. If the macro is currently being executed, an ERROR diagnostic message is issued.

%SYSMSTORECLEAR Statement

Closes the stored compiled macro catalog associated with the libref specified in the SASMSTORE= option and clears the libref.

Type: Macro statement

Restriction: Allowed in macro definition and open code

See: SASMSTORE= system option

Syntax

%SYSMSTORECLEAR;

Details

Use the %SYSMSTORECLEAR statement to close the stored compiled macro catalog and to clear the previous libref when switching between SASMSTORE= libraries.

Note: If any stored compiled macro from the library specified by the SASMSTORE= system option is still executing, the following will occur:

- an ERROR diagnostic message will be issued
- the library will not be closed
- the libref will not be cleared

%SYSRPUT Statement

Assigns the value of a macro variable on a remote host to a macro variable on the local host.

Type: Macro statement

Restriction: Allowed in macro definitions or open code

Requirement: SAS/CONNECT

See: "SYSERR Automatic Macro Variable" on page 200, "SYSINFO Automatic Macro Variable" on page 204, and "%SYSLPUT Statement" on page 317

Syntax

%SYSRPUT *local-macro-variable=remote-macro-variable*;

Required Arguments

local-macro-variable
is the name of a macro variable with no leading ampersand or a text expression that produces the name of a macro variable. This name must be a macro variable stored on the local host.

remote-macro-variable
is the name of a macro variable with no leading ampersand or a text expression that produces the name of a macro variable. This name must be a macro variable stored on a remote host.

Details

The %SYSRPUT statement is submitted with SAS/CONNECT to a remote host to retrieve the value of a macro variable stored on the remote host. %SYSRPUT assigns that value to a macro variable on the local host. %SYSRPUT is similar to the %LET macro statement because it assigns a value to a macro variable. However, %SYSRPUT assigns a value to a variable on the local host, not on the remote host where the statement is processed. The %SYSRPUT statement places the macro variable into the global symbol table in the client session.

Note: The names of the macro variables on the remote and local hosts must not contain a leading ampersand.

The %SYSRPUT statement is useful for capturing the value of the automatic macro variable SYSINFO and passing that value to the local host. SYSINFO contains return-code information provided by some SAS procedures. Both the UPLOAD and the DOWNLOAD procedures of SAS/CONNECT can update the macro variable SYSINFO and set it to a nonzero value when the procedure terminates due to errors. You can use %SYSRPUT on the remote host to send the value of the SYSINFO macro variable back to the local SAS session. Thus, you can submit a job to the remote host and test whether a PROC UPLOAD or DOWNLOAD step has successfully completed before beginning another step on either the remote host or the local host.

For details about using %SYSRPUT, see the documentation for SAS/CONNECT Software.

To create a new macro variable or modify the value of an existing macro variable on a remote host or server, use the %SYSLPUT macro statement.

Example: Checking the Value of a Return Code on a Remote Host

This example illustrates how to download a file and return information about the success of the step from a noninteractive job. When remote processing is completed, the job then checks the value of the return code stored in RETCODE. Processing continues on the local host if the remote processing is successful.

The %SYSRPUT statement is useful for capturing the value returned in the SYSINFO macro variable and passing that value to the local host. The SYSINFO macro variable contains return-code information provided by SAS procedures. In the example, the %SYSRPUT statement follows a PROC DOWNLOAD step, so the value returned by SYSINFO indicates the success of the PROC DOWNLOAD step:

```
rsubmit;
   %macro download;
      proc download data=remote.mydata out=local.mydata;
      run;
```

```
      %sysrput retcode=&sysinfo;
   %mend download;
   %download
endrsubmit;
%macro checkit;
   %if &retcode = 0 %then %do;
      further processing on local host
   %end;
%mend checkit;
%checkit
```

A SAS/CONNECT batch (noninteractive) job always returns a system condition code of 0. To determine the success or failure of the SAS/CONNECT noninteractive job, use the %SYSRPUT macro statement to check the value of the automatic macro variable SYSERR. To determine what remote system the SAS/CONNECT conversation is attached to, remote submit the following statement:

```
%sysrput rhost=&sysscp;
```

%WINDOW Statement

Defines customized windows.

Type: Macro statement

Restriction: Allowed in macro definitions or open code

See: "%DISPLAY Statement" on page 287 and "%INPUT Statement" on page 299

Syntax

%WINDOW*window-name*<*window-option-1* <...*window-option-n*> *group-definition-1* <...*group-definition-n*>> *field-definition-1* <...*field-definition-n*>;

Required Arguments

window-name
names the window. *Window-name* must be a SAS name.

window-option-1 <...window-option-n>
specifies the characteristics of the window as a whole. Specify all window options before any field or group definitions. These window options are available:

COLOR=*color*
specifies the color of the window background. The default color of the window and the contents of its fields are both device-dependent. *Color* can be one of these:

BLACK

BLUE

BROWN

CYAN

GRAY (or GREY)

GREEN

MAGENTA

ORANGE

PINK

RED

WHITE

YELLOW

Operating Environment Information
The representation of colors might vary, depending on the display device that you use. In addition, on some display devices the background color affects the entire window; on other display devices, it affects only the window border.

COLUMNS=*columns*
specifies the number of display columns in the window, including borders. A window can contain any number of columns and can extend beyond the border of the display. This feature is useful when you need to display a window on a device larger than the one on which you developed it. By default, the window fills all remaining columns in the display.

Operating Environment Information
The number of columns available depends on the type of display device that you use. Also, the left and right borders each use from 0 to 3 columns on the display depending on your display device. If you create windows for display on different types of display devices, ensure that all fields can be displayed in the narrowest window.

ICOLUMN=*column*
specifies the initial column within the display at which the window is displayed. By default, the macro processor begins the window at column 1 of the display.

IROW=*row*
specifies the initial row (line) within the display at which the window is displayed. By default, the macro processor begins the window at row 1 of the display.

KEYS=<<*libref.*>*catalog.*>*keys-entry*
specifies the name of a KEYS catalog entry that contains the function key definitions for the window. If you omit *libref* and *catalog*, SAS uses SASUSER.PROFILE.*keys-entry*.

If you omit the KEYS= option, SAS uses the current function key settings defined in the KEYS window.

MENU=<<*libref.*>*catalog.*>*pmenu-entry*
specifies the name of a menu that you have built with the PMENU procedure. If you omit *libref* and *catalog*, SAS uses SASUSER.PROFILE.*pmenu-entry*.

ROWS=*rows*
specifies the number of rows in the window, including borders. A window can contain any number of rows and can extend beyond the border of the display device. This feature is useful when you need to display a window on a device larger than the one on which you developed it. If you omit a number, the window fills all remaining rows in the display device.

Operating Environment Information
The number of rows available depends on the type of display device that you use.

group-definition
names a group and defines all fields within a group. The form of *group definition* is GROUP=*group field-definition* <*. . . field-definition-n*> where *group* names a group

of fields that you want to display in the window collectively. A window can contain any number of groups of fields. If you omit the GROUP= option, the window contains one unnamed group of fields. *Group* must be a SAS name.

Organizing fields into groups enables you to create a single window with several possible contents. To refer to a particular group, use *window.group*

field-definition

identifies and describes a macro variable or string that you want to display in the window. A window can contain any number of fields.

You use a field to identify a macro variable value (or constant text) to be displayed, its position within the window, and its attributes. Enclose constant text in quotation marks. The position of a field is determined by beginning row and column. The attributes that you can specify include color, whether you can enter a value into the field, and characteristics such as highlighting.

The form of a field definition containing a macro variable is

<row> <column> macro-variable<field-length> <options>

The form of a field definition containing constant text is

<row> <column>'text' | "text"<options>

The elements of a field definition are

row

specifics the row (line) on which the macro variable or constant text is displayed. Each row specification consists of a pointer control and, usually, a macro expression that generates a number. These row pointer controls are available:

#macro-expression

specifies the row within the window given by the value of the macro expression. The macro expression must either be a positive integer or generate a positive integer.

/ (forward slash)

moves the pointer to column 1 of the next line.

The macro processor evaluates the macro expression when it defines the window, not when it displays the window. Thus, the row position of a field is fixed when the field is being displayed.

If you omit *row* in the first field of a group, the macro processor uses the first line of the window. If you omit *row* in a later field specification, the macro processor continues on the line from the previous field.

The macro processor treats the first usable line of the window as row 1 (that is, it excludes the border, command line or menu bar, and message line).

Specify either *row* or *column* first.

column

specifies the column in which the macro variable or constant text begins. Each column specification consists of a pointer control and, usually, a macro expression that generates a number. These column pointer controls are available:

@macro-expression

specifies the column within the window given by the value of the macro expression. The macro expression must either be a positive integer or generate a positive integer.

+macro-expression
moves the pointer the number of columns given by the value of the macro expression. The macro expression must either be a positive integer or generate a positive integer.

The macro processor evaluates the macro expression when it defines the window, not when it displays the window. Thus, the column position of a field is fixed when the field is being displayed.

The macro processor treats the column after the left border as column 1. If you omit *column*, the macro processor uses column 1.

Specify either *column* or *row* first.

macro-variable
names a macro variable to be displayed or to receive the value that you enter at that position. The macro variable must either be a macro variable name (not a macro variable reference) or it must be a macro expression that generates a macro variable name.

By default, you can enter or change a macro variable value when the window containing the value is displayed. To display the value without changes, use the PROTECT= option.

CAUTION:
Do not overlap fields. Do not let a field overlap another field displayed at the same time. Unexpected results, including the incorrect assignment of values to macro variables, might occur. (Some display devices treat adjacent fields with no intervening blanks as overlapping fields.) SAS writes a warning in the SAS log if fields overlap.

field-length
is an integer specifying how many positions in the current row are available for displaying the macro variable's value or for accepting input. The maximum value of *field-length* is the number of positions remaining in the row. You cannot extend a field beyond one row.

Note: The field length does not affect the length stored for the macro variable. The field length affects only the number of characters displayed or accepted for input in a particular field.

If you omit *field-length* when the field contains an existing macro variable, the macro processor uses a field equal to the current length of the macro variable value. This value can be up to the number of positions remaining in the row or remaining until the next field begins.

CAUTION:
Specify a field length whenever a field contains a macro variable. If the current value of the macro variable is null, as in a macro variable defined in a %GLOBAL or %LOCAL statement, the macro processor uses a field length of 0. You cannot input any characters into the field.

If you omit *field-length* when the macro variable is created in that field, the macro processor uses a field length of zero. Specify a field length whenever a field contains a macro variable.

'text' | "text"
contains constant text to be displayed. The text must be enclosed in either single or double quotation marks. You cannot enter a value into a field containing constant text.

options
can include the following:

ATTR=*attribute* | (*attribute-1* <. . . , *attribute-n*>) A=*attribute* | (*attribute-1* <. . . , *attribute-n*>)
controls several display attributes of the field. The display attributes and combinations of display attributes available depend on the type of display device that you use.

BLINK	causes the field to blink.
HIGHLIGHT	displays the field at high intensity.
REV_VIDEO	displays the field in reverse video.
UNDERLINE	underlines the field.

AUTOSKIP=YES | NO AUTO=YES | NO
controls whether the cursor moves to the next unprotected field of the current window or group when you have entered data in all positions of a field. If you specify AUTOSKIP=YES, the cursor moves automatically to the next unprotected field. If you specify AUTOSKIP=NO, the cursor does not move automatically. The default is AUTOSKIP=YES.

COLOR=*color* C=*color*
specifies a color for the field. The default color is device-dependent. *Color* can be one of these:

BLACK
BLUE
BROWN
CYAN
GRAY (or GREY)
GREEN
MAGENTA
ORANGE
PINK
WHITE
YELLOW

DISPLAY=YES | NO
determines whether the macro processor displays the characters that you are entering into a macro variable value as you enter them. If you specify DISPLAY=YES (the default value), the macro processor displays the characters as you enter them. If you specify DISPLAY=NO, the macro processor does not display the characters as you enter them.

DISPLAY=NO is useful for applications that require users to enter confidential information, such as passwords. Use the DISPLAY= option only with fields containing macro variables; constant text is displayed automatically.

PROTECT=YES | NO P=YES | NO
controls whether information can be entered into a field containing a macro variable. If you specify PROTECT=NO (the default value), you can enter information. If you specify PROTECT=YES, you cannot enter information

into a field. Use the PROTECT= option only for fields containing macro variables; fields containing text are automatically protected.

REQUIRED=YES | NO
: determines whether you must enter a value for the macro variable in that field. If you specify REQUIRED=YES, you must enter a value into that field in order to remove the display from the window. You cannot enter a null value into a required field. If you specify REQUIRED=NO (the default value), you do not have to enter a value in that field in order to remove the display from the window. Entering a command on the command line of the window removes the effect of REQUIRED=YES.

Details

Use the %WINDOW statement to define customized windows that are controlled by the macro processor. These windows have command and message lines. You can use these windows to display text and accept input. In addition, you can invoke windowing environment commands, assign function keys, and use a menu generated by the PMENU facility.

You must define a window before you can display it. The %WINDOW statement defines macro windows; the %DISPLAY statement displays macro windows. Once defined, a macro window exists until the end of the SAS session, and you can display a window or redefine it at any point.

Defining a macro window within a macro definition causes the macro processor to redefine the window each time the macro executes. If you repeatedly display a window whose definition does not change, it is more efficient to do one of the following:

- define the window outside a macro
- define the window in a macro that you execute once rather than in the macro in which you display it

If a %WINDOW statement contains the name of a new macro variable, the macro processor creates that variable with the current scope. The %WINDOW statement creates two automatic macro variables.

SYSCMD
: contains the last command from the window's command line that was not recognized by the windowing environment.

SYSMSG
: contains text that you specify to be displayed on the message line.

Note: Windowing environment file management, scrolling, searching, and editing commands are not available to macro windows.

Examples

Example 1: Creating an Application Welcome Window

This %WINDOW statement creates a window with a single group of fields:

```
%window welcome color=white
          #5 @28 'Welcome to SAS.' attr=highlight
             color=blue
          #7 @15
             "You are executing Release &sysver on &sysday, &sysdate.."
          #12 @29 'Press ENTER to continue.';
```

The WELCOME window fills the entire display device. The window is white, the first line of text is blue, and the other two lines are black at normal intensity. The WELCOME window does not require you to input any values. However, you must press ENTER to remove the display from the window.

Note: Two periods are a needed delimiter for the reference to the macro variables SYSVER, SYSDAY, and SYSDATE.

Example 2: Creating Macro Variables from Input Information

The following example prompts for information and creates macro variables from that information:

```
%window info
  #5 @5 'Please enter userid:'
  #5 @26 id 8 attr=underline
  #7 @5 'Please enter password:'
  #7 @28 pass 8 attr=underline display=no;
%display info;
%put userid entered was &id;
%put password entered was &pass;
```

Chapter 20
System Options for Macros

System Options for Macros

There are several SAS system options that apply to the macro facility.

Dictionary

CMDMAC System Option

Controls command-style macro invocation.

Valid in:	Configuration file OPTIONS windowOPTIONS statement SAS invocation
PROC OPTIONS GROUP=	MACRO
Type:	System option
Default:	NOCMDMAC

Syntax

CMDMAC | NOCMDMAC

Required Arguments

CMDMAC

specifies that the macro processor examine the first word of every windowing environment command to see whether it is a command-style macro invocation.

Note: When CMDMAC is in effect, SAS searches the macro libraries first and executes any member it finds with the same name as the first word in the windowing environment command that was issued. Unexpected results can occur.

NOCMDMAC

specifies that no check be made for command-style macro invocations. If the macro processor encounters a command-style macro call when NOCMDMAC is in effect, it treats the call as a SAS command and produces an error message if the command is not valid or is not used correctly.

Details

The CMDMAC system option controls whether macros defined as command-style macros can be invoked with command-style macro calls or if these macros must be invoked with name-style macro calls. These two examples illustrate command-style and name-style macro calls, respectively:

- `macro-name parameter-value-1 parameter-value-2`
- `%macro-name(parameter-value-1, parameter-value-2)`

When you use CMDMAC, processing time is increased because the macro facility searches the macros compiled during the current session for a name corresponding to the first word on the command line. If the MSTORED option is in effect, the libraries containing compiled stored macros are searched for a name corresponding to that word. If the MAUTOSOURCE option is in effect, the autocall libraries are searched for a name corresponding to that word. If the MRECALL system option is also in effect, processing time can be increased further because the search continues even if a word was not found in a previous search.

Regardless of which option is in effect, you can use a name-style invocation to call any macro, including command-style macros.

Comparisons

Name-style macros are the more efficient choice for invoking macros because the macro processor searches only for a macro name corresponding to a word following a percent sign.

IMPLMAC System Option

Controls statement-style macro invocation.

Valid in: Configuration file OPTIONS windowOPTIONS statement SAS invocation

PROC OPTIONS GROUP= MACRO

Type: System option

Default: NOIMPLMAC

Syntax

IMPLMAC | NOIMPLMAC

Required Arguments

IMPLMAC
specifies that the macro processor examine the first word of every submitted statement to see whether it is a statement-style macro invocation.

Note: When IMPLMAC is in effect, SAS searches the macro libraries first and executes any macro it finds with the same name as the first word in the SAS statement that was submitted. Unexpected results can occur.

NOIMPLMAC
specifies that no check be made for statement-style macro invocations. This is the default. If the macro processor encounters a statement-style macro call when NOIMPLMAC is in effect, it treats the call as a SAS statement. SAS produces an error message if the statement is not valid or if it is not used correctly.

Details

The IMPLMAC system option controls whether macros defined as statement-style macros can be invoked with statement-style macro calls or if these macros must be invoked with name-style macro calls. These examples illustrate statement-style and name-style macro calls, respectively:

- `macro-name parameter-value-1  parameter-value-2;`
- `%macro-name(parameter-value-1, parameter-value-2)`

When you use IMPLMAC, processing time is increased because SAS searches the macros compiled during the current session for a name corresponding to the first word of each SAS statement. If the MSTORED option is in effect, the libraries containing compiled stored macros are searched for a name corresponding to that word. If the MAUTOSOURCE option is in effect, the autocall libraries are searched for a name corresponding to that word. If the MRECALL system option is also in effect, processing time can be increased further because the search continues even if a word was not found in a previous search.

Regardless of which option is in effect, you can call any macro with a name-style invocation, including statement-style macros.

Note: If a member in an autocall library or stored compiled macro catalog has the same name as an existing windowing environment command, SAS searches for the macro first if CMDMAC is in effect. Unexpected results can occur.

Comparisons

Name-style macros are a more efficient choice to use when you invoke macros because the macro processor searches only for the macro name that corresponds to a word that follows a percent sign.

MACRO System Option

Controls whether the SAS macro language is available.

Valid in: Configuration file SAS invocation

PROC OPTIONS GROUP= MACRO

Type: System option

Default: MACRO

Syntax

MACRO | NOMACRO

Required Arguments

MACRO
enables SAS to recognize and process macro language statements, macro calls, and macro variable references.

NOMACRO
prevents SAS from recognizing and processing macro language statements, macro calls, and macro variable references. The item generally is not recognized, and an error message is issued. If the macro facility is not used in a job, a small performance gain can be made by setting NOMACRO because there is no overhead of checking for macros or macro variables.

MAUTOCOMPLOC System Option

Displays in the SAS log the source location of an autocall macro when the autocall macro is compiled.

Valid in: Configuration file, OPTIONS window, OPTIONS statement, SAS invocation

PROC OPTIONS GROUP= MACRO

Type: System option

Default: NOMAUTOCOMPLOC

Syntax

MAUTOCOMPLOC|NOMAUTOCOMPLOC

Required Arguments

MAUTOCOMPLOC
displays the autocall macro source location in the SAS log when the autocall macro is compiled.

NOMAUTOCOMPLOC
prevents the autocall macro source location from being written to the SAS log.

Details

The display created by the MAUTOCOMPLOC system option of the autocall macro source location in the log is not affected by either the MAUTOLOCDISPLAY or the MLOGIC system options.

MAUTOLOCDISPLAY System Option

Specifies whether to display the source location of the autocall macros in the log when the autocall macro is invoked.

Valid in: Configuration fileOPTIONS windowOPTIONS statementSAS invocation

PROC OPTIONS GROUP= MACRO

Type: System option

Default: NOMAUTOLOCDISPLAY

Syntax

MAUTOLOCDISPLAY | NOMAUTOLOCDISPLAY

Required Arguments

MAUTOLOCDISPLAY
enables MACRO to display the autocall macro source location in the log when the autocall macro is invoked.

NOMAUTOLOCDISPLAY
prevents the autocall macro source location from being displayed in the log when the autocall macro is invoked. NOMAUTOLOCDISPLAY is the default.

Details

When both MAUTOLOCDISPLAY and MLOGIC options are set, only the MLOGIC listing of the autocall source location is displayed.

MAUTOLOCINDES System Option

Specifies whether the macro processor prepends the full pathname of the autocall source file to the description field of the catalog entry of compiled autocall macro definition in the WORK.SASMACR catalog.

Valid in: Configuration file, OPTIONS window, OPTIONS statement, SAS invocation

PROC OPTIONS GROUP= MACRO

Type: System option

Default: NOMAUTOLOCINDES

See: SAS log

Syntax

MAUTOLOCINDES|NOMAUTOLOCINDES

Required Arguments

MAUTOLOCINDES
causes the macro processor to prepend the full pathname of the autocall macro source file to the description field of the catalog entry of the compiled autocall macro definition in the WORK.SASMACR catalog.

NOMAUTOLOCINDES
no changes to the description field autocall macro definitions in the WORK.SASMACR catalog.

Details

Use MAUTOLOCINDES to help determine where autocall macro definition source code is located. The following is an example that shows the output that contains the full pathname:

```
  options mautolocindes;
  %put %lowcase(THIS);

this

  proc catalog cat=work.sasmacr;contents;run;

                       Contents of Catalog WORK.SASMACR

# Name    Type          Create Date       Modified Date Description

1 LOWCASE MACRO 12Sep10:10:36:57    12Sep10:10:36:57    C:\SASv9\sas\dev\
                                                        mva-v930\shell\auto\
```

MAUTOSOURCE System Option

Specifies whether the autocall feature is available.

Valid in: Configuration fileOPTIONS windowOPTIONS statementSAS invocation

PROC OPTIONS GROUP= MACRO

Type: System option

Default: MAUTOSOURCE

Syntax

MAUTOSOURCE | NOMAUTOSOURCE

Required Arguments

MAUTOSOURCE
causes the macro processor to search the autocall libraries for a member with the requested name when a macro name is not found in the WORK library.

NOMAUTOSOURCE
prevents the macro processor from searching the autocall libraries when a macro name is not found in the WORK library.

Details

When the macro facility searches for macros, it searches first for macros compiled in the current SAS session. If the MSTORED option is in effect, the macro facility next searches the libraries containing compiled stored macros. If the MAUTOSOURCE option is in effect, the macro facility next searches the autocall libraries.

MCOMPILENOTE System Option

Issues a NOTE to the SAS log. The note contains the size and number of instructions upon the completion of the compilation of a macro.

Valid in: Configuration fileOPTIONS windowOPTIONS StatementSAS invocation

PROC OPTIONS GROUP= MACRO

Type: System option

Default: NONE

Syntax

MCOMPILENOTE=<NONE | NOAUTOCALL | ALL>

Required Arguments

NONE
prevents any NOTE from being written to the log.

NOAUTOCALL
prevents any NOTE from being written to the log for AUTOCALL macros, but does issue a NOTE to the log upon the completion of the compilation of any other macro.

ALL
issues a NOTE to the log. The note contains the size and number of instructions upon the completion of the compilation of any macro.

Details

The NOTE confirms that the compilation of the macro was completed. When the option is on and the NOTE is issued, the compiled version of the macro is available for execution. A macro can successfully compile, but still contain errors or warnings that will cause the macro to not execute as you intended.

Example: Using MCOMPILENOTE System Option

A macro can actually compile and still contain errors. Here is an example of the NOTE without errors:

```
option mcompilenote=noautocall;
%macro mymacro;
%mend mymacro;
```

Output to the log:

```
NOTE: The macro MYMACRO completed compilation without errors.
```

Here is an example of the NOTE with errors:

```
%macro yourmacro;
%end;
%mend yourmacro;
```

Output to the log:

```
ERROR: There is no matching %DO statement for the %END statement.
       This statement will be ignored.
NOTE: The macro YOURMACRO completed compilation with errors.
```

MCOMPILE System Option

Specifies whether to allow new definitions of macros.

Valid in: Configuration fileOPTIONS windowOPTIONS statementSAS invocation

PROC OPTIONS GROUP= MACRO

Type: System option

Default: MCOMPILE

Syntax

MCOMPILE | NOMCOMPILE

Required Arguments

MCOMPILE
allows new macro definitions.

NOMCOMPILE
disallows new macro definitions.

Details

The MCOMPILE system option allows new definitions of macros.

The NOMCOMPILE system option prevents new definitions of macros. It does not prevent the use of existing stored compiled or autocall macros.

MCOVERAGE System Options

Enables the generation of coverage analysis data.

Valid in:	Configuration file, OPTIONS window, OPTIONS statement, SAS invocation
PROC OPTIONS GROUP=	Macro
Type:	System option
Default:	NOMCOVERAGE
Requirement:	Must use MCOVERAGELOC= system option

Syntax

MCOVERAGE|NOMCOVERAGE

Required Arguments

MCOVERAGE
enables the generation of coverage analysis data.

NOMCOVERAGE
prevents the generation of coverage analysis data.

Details

MCOVERAGE system option controls the generation of *coverage analysis data*, which is information needed to ensure proper testing of SAS Solutions products before their release.

The format of the coverage analysis data is a space delimited flat text file that contains three types of records. Each record begins with a numeric record type. The line numbers in the data are relative line numbers based on the %MACRO keyword used to define the macro. You must use the MCOVERAGELOC= system option to specify the location of the coverage analysis data file. See "MCOVERAGELOC= System Option" on page 340.

Note: Because nested macro definitions are stored as model text with line breaks collapsed, it is recommended that nested macro definitions not be used in macro definitions that will later be analyzed for execution coverage.

Below are explanations for each of the three record types.

Record type 1:

`1 n n macroname`

1
: record type

`n`
: first line number

`n`
: last line number

`macroname`
: macro name

Record type 1 indicates the beginning of the execution of a macro. Record type 1 appears once for each invocation of a macro.

Record type 2:

`2 n n macroname`

2
: record type

`n`
: first line number

`n`
: last line number

`macroname`
: macro name

Record type 2 indicates the lines of a macro that have executed. A single line of a macro might cause more than one record to be generated.

Record type 3:

`3 n n macroname`

3
: record type

`n`
: first line number

`n`
: last line number

`macroname`
: macro name

Record type 3 indicates which lines of the macro cannot be executed because no code was generated from them. These lines might be either commentary lines or lines that cause no macro code to be generated.

The following is a sample program log:

```
Sample Program Log:
NOTE: Copyright (c) 2002-2008 by SAS Institute Inc., Cary, NC, USA.
```

```
NOTE: SAS (r) Proprietary Software 9.3 (TS1B0)
      Licensed to SAS Institute Inc., Site 1.
NOTE: This session is executing on the XP_PRO  platform.



NOTE: SAS initialization used:
      real time           0.45 seconds
      cpu time            0.20 seconds

1          options source source2;
2
3          options mcoverage mcoverageloc='./foo.dat';
4
5          /*  1 */ %macro
6          /*  2 */ foo (
7          /*  3 */ arg,
8          /*  4 */
9          /*  5 */
10         /*  6 */ arg2
11         /*  7 */
12         /*  8 */
13         /*  9 */ =
14         /* 10 */
15         /* 11 */ This is the default value of arg2)
16         /* 12 */ ;
17         /* 13 */ /*  This is a number of lines of comments     */
18         /* 14 */ /*  which presumably will help the maintainer */
19         /* 15 */ /*  of this macro to know what to do to keep  */
20         /* 16 */ /*  this silly piece of code current          */
21         /* 17 */   %if &arg %then %do;
22         /* 18 */      data _null_;
23         /* 19 */      x=1;
24         /* 20 */   %end;
25         /* 21 */ %* this is a macro comment statement
26         /* 22 */    that also can be used to document features
27         /* 23 */    and other stuff about the macro;
28         /* 24 */   %else
29         /* 25 */   %do;
30         /* 26 */      DATA _NULL_;
31         /* 27 */      y=1;
32         /* 28 */   %end;
33         /* 29 */      run;
34         /* 30 */
35         /* 31 */
36         /* 32 */
37         /* 33 */
38         /* 34 */
39         /* 35 */ %mend
40         /* 36 */
41         /* 37 */
42         /* 38 */
43         /* 39 */
44         /* 40 */
45         /* 41 */
46         /* 42 */
```

```
47          /* 43 */ foo This is text which should generate a warning!  ;
WARNING: Extraneous information on %MEND statement ignored for macro
definition FOO.
```

MCOVERAGELOC= System Option

Specifies the location of the coverage analysis data file.

Valid in: Configuration file, OPTIONS window, OPTIONS statement, SAS invocation

PROC OPTIONS GROUP= Macro

Type: System option

Requirement: Use with MCOVERAGE system option

See: "MCOVERAGE System Options" on page 337

Syntax

MCOVERAGELOC=*fileref|file_specification*

Required Argument

fileref|file_specification
a SAS fileref or an external file specification enclosed in quotation marks.

Details

This MCOVERAGELOC = system option specifies where the coverage analysis is to be written. The option takes either an external file specification enclosed in quotation marks or a SAS fileref.

MERROR System Option

Specifies whether the macro processor issues a warning message when a macro reference cannot be resolved.

Valid in: Configuration fileOPTIONS windowOPTIONS statementSAS invocation

PROC OPTIONS GROUP= MACRO

Type: System option

Default: MERROR

Syntax

MERROR | NOMERROR

Required Arguments

MERROR
issues the following warning message when the macro processor cannot match a macro reference to a compiled macro:

```
WARNING: Apparent invocation of macro %text not resolved.
```

NOMERROR
issues no warning messages when the macro processor cannot match a macro reference to a compiled macro.

Details

Several conditions can prevent a macro reference from resolving. These conditions appear when

- a macro name is misspelled
- a macro is called before being defined
- strings containing percent signs are encountered. For example:

```
TITLE Cost Expressed as %Sales;
```

If your program contains a percent sign in a string that could be mistaken for a macro keyword, specify NOMERROR.

MEXECNOTE System Option

Specifies whether to display macro execution information in the SAS log at macro invocation.

Valid in:	Configuration fileOPTIONS windowOPTIONS statementSAS invocation
PROC OPTIONS GROUP=	MACRO
Type:	System option
Default:	NOMEXECNOTE
See:	MEXECSIZE on page 342

Syntax

MEXECNOTE | NOMEXECNOTE

Required Arguments

MEXECNOTE
displays the macro execution information in the log when the macro is invoked.

NOMEXECNOTE
does not display the macro execution information in the log when the macro is invoked.

Details

The MEXECNOTE option controls the generation of a NOTE in the SAS log that indicates the macro execution mode.

MEXECSIZE System Option

Specifies the maximum macro size that can be executed in memory.

Valid in: Configuration fileOPTIONS windowOPTIONS statementSAS invocation

PROC OPTIONS GROUP= MACRO

Type: System option

Default: 65536

See: MEXECNOTE on page 341 and MCOMPILENOTE on page 335

Syntax

MEXECSIZE=*n* | *n*K | *n*M | *n*G | *n*T | *hex*X | MIN | MAX

Required Arguments

n
specifies the maximum size macro to be executed in memory available in bytes.

***n*K**
specifies the maximum size macro to be executed in memory available in kilobytes.

***n*M**
specifies the maximum size macro to be executed in memory available in megabytes.

***n*G**
specifies the maximum size macro to be executed in memory available in gigabytes.

***n*T**
specifies the maximum size macro to be executed in memory available in terabytes.

MIN
specifies the minimum size macro to be executed in memory. Minimum value is 0.

MAX
specifies the maximum size macro to be executed in memory. Maximum value is

***hex*X**
specifies the maximum size macro to be executed in memory by a hexadecimal number followed by an X.

Details

Use the MEXECSIZE option to control the maximum size macro that will be executed in memory as opposed to being executed from a file. The MEXECSIZE option value is the compiled size of the macro. Memory is allocated only when the macro is executed. After the macro completes, the memory is released. If memory is not available to execute the macro, an out-of-memory message is written to the SAS log. Use the MCOMPILENOTE option to write to the SAS log the size of the compiled macro. The MEMSIZE option does not affect the MEXECSIZE option.

MFILE System Option

Specifies whether MPRINT output is routed to an external file.

Valid in: Configuration fileOPTIONS window OPTIONS statementSAS invocation

PROC OPTIONS GROUP= MACRO

Type: System option

Default: NOMFILE

Requirement: MPRINT option

See: "MPRINT System Option" on page 349

Syntax

MFILE | NOMFILE

Required Arguments

MFILE
routes output produced by the MPRINT option to an external file. This option is useful for debugging.

NOMFILE
does not route MPRINT output to an external file.

Details

The MPRINT option must also be in effect to use MFILE, and an external file must be assigned the fileref MPRINT. Macro-generated code that is displayed by the MPRINT option in the SAS log during macro execution is written to the external file referenced by the fileref MPRINT.

If MPRINT is not assigned as a fileref or if the file cannot be accessed, warnings are written to the SAS log and MFILE is set to off. To use the feature again, you must specify MFILE again and assign the fileref MPRINT to a file that can be accessed.

MINDELIMITER= System Option

Specifies the character to be used as the delimiter for the macro IN operator.

Valid in: Configuration fileOPTIONS windowOPTIONS statementSAS invocation

PROC OPTIONS GROUP= MACRO

Type: System option

Default: a blank

See: "MINOPERATOR System Option" on page 345 and "%MACRO Statement" on page 304

Syntax

MINDELIMITER=<"*option*">

Required Argument

option

is a character enclosed in double or single quotation marks. The character will be used as the delimiter for the macro IN operator. Here is an example:double quotation marks

```
mindelimiter=",";
```

or single quotation marks

```
mindelimiter=',';
```

Details

The option value is retained in original case and can have a maximum length of one character. The default value of the MINDELIMITER option is a blank.

You can use the **#** character instead of IN.

Note: When the IN or **#** operator is used in a macro, the delimiter that is used at the execution time of the macro is the value of the MINDELIMITER option at the time of the compilation of the macro. A specific delimiter value for use during the execution of the macro other than the current value of the MINDELIMITER system option might be specified on the macro definition statement:

```
%macro macroname / mindelimiter=',';
```

Comparisons

The following is an example using a specified delimiter in an IN operator:

```
%put %eval(a in d,e,f,a,b,c); /* should print 0 */
%put %eval(a in d e f a b c); /* should print 1 */
option mindelimiter=',';
%put %eval(a in d,e,f,a,b,c); /* should print 1 */
%put %eval(a in d e f a b c); /* should print 0 */
```

The following is the output to the SAS log:

```
NOTE: Copyright (c) 2007-2008 by SAS Institute Inc., Cary, NC, USA.
NOTE: SAS (r) Proprietary Software Version 9.2 (TS A0)
      Licensed to SAS Institute Inc., Site 0000000001.
NOTE: This session is executing on the WIN_NT  platform.
NOTE: SAS initialization used:
      real time           1.02 seconds
      cpu time            0.63 seconds
  %put %eval(a in d,e,f,a,b,c); /* should print 0 */
0
  %put %eval(a in d e f a b c); /* should print 1 */
1
  option mindelimiter=',';
  %put %eval(a in d,e,f,a,b,c); /* should print 1 */
1
  %put %eval(a in d e f a b c); /* should print 0 */
0
```

MINOPERATOR System Option

Controls whether the macro processor recognizes and evaluates the IN (#) logical operator

Valid in:	Configuration fileOPTIONS windowOPTIONS statementSAS invocation
PROC OPTIONS GROUP=	MACRO
Type:	System option
Default:	NOMINOPERATOR

Syntax

MINOPERATOR | NOMINOPERATOR

Required Arguments

MINOPERATOR
causes the macro processor to recognize and evaluate both the mnemonic operator **IN** or the special character **#** as a logical operator in expressions.

NOMINOPERATOR
causes the macro processor to recognize both the mnemonic operator **IN** and the special character **#** as regular characters.

Details

Use the MINOPERATOR system option or in the %MACRO statement if you want to use the **IN** (**#**) as operators in expressions:

```
options minoperator;
```

To use IN or # as operators in expressions evaluated during the execution of a specific macro, use the MINOPERATOR keyword on the definition of the macro:

```
%macro macroname / minoperator;
```

The macro IN operator is similar to the DATA step IN operator, but not identical. The following is a list of differences:

- The macro IN operator cannot search a numeric array.
- The macro IN operator cannot search a character array.
- A colon (**:**) is not recognized as a shorthand notation to specify a range, such as **1:10** means 1 through 10. Instead, you use the following in a macro:

  ```
  %eval(3 in 1 2 3 4 5 6 7 8 9 10);
  ```

- The default delimiter for list elements is a blank. For more information, see “MINDELIMITER= System Option” on page 343.
- Both operands must contain a value.

  ```
  %put %eval(a IN a b c d); /*Both operands are present. */
  ```

  If an operand contains a null value, an error is generated.

  ```
  %put %eval(  IN a b c d); /*Missing first operand. */
  ```

or

```
%put %eval(a IN); /*Missing second operand. */
```

Whether the first or second operand contains a null value, the same error is written to the SAS log:

```
ERROR: Operand missing for IN operator in argument to %EVAL function.
```

The following example uses the macro IN operator to search a character string:

```
%if &state in (NY NJ PA) %then %let &region = %eval(&region + 1);
```

For more information, see "Defining Arithmetic and Logical Expressions " on page 72.

MLOGIC System Option

Specifies whether the macro processor traces its execution for debugging.

Valid in: Configuration fileOPTIONS windowOPTIONS statementSAS invocation

PROC OPTIONS GROUP= MACRO
LOGCONTROL

Type: System option

Default: NOMLOGIC

See: "The SAS Log" in Chapter 9 of *SAS Language Reference: Concepts*

Syntax

MLOGIC | NOMLOGIC

Required Arguments

MLOGIC
causes the macro processor to trace its execution and to write the trace information to the SAS log. This option is a useful debugging tool.

NOMLOGIC
does not trace execution. Use this option unless you are debugging macros.

Details

Use MLOGIC to debug macros. Each line generated by the MLOGIC option is identified with the prefix MLOGIC(*macro-name*):. If MLOGIC is in effect and the macro processor encounters a macro invocation, the macro processor displays messages that identify the following:

- the beginning of macro execution
- values of macro parameters at invocation
- execution of each macro program statement
- whether each %IF condition is true or false
- the ending of macro execution

Note: Using MLOGIC can produce a great deal of output.

For more information about macro debugging, see "Macro Facility Error Messages and Debugging" on page 120.

Example: Tracing Macro Execution

In this example, MLOGIC traces the execution of the macros MKTITLE and RUNPLOT:

```
%macro mktitle(proc,data);
    title "%upcase(&proc) of %upcase(&data)";
%mend mktitle;
%macro runplot(ds);
   %if %sysprod(graph)=1 %then
      %do;
         %mktitle (gplot,&ds)
         proc gplot data=&ds;
            plot style*price
                / haxis=0 to 150000 by 50000;
         run;
         quit;
      %end;
   %else
      %do;
         %mktitle (plot,&ds)
         proc plot data=&ds;
            plot style*price;
         run;
         quit;
      %end;
%mend runplot;
options mlogic;
%runplot(sasuser.houses)
```

When this program executes, this MLOGIC output is written to the SAS log:

```
MLOGIC(RUNPLOT):  Beginning execution.
MLOGIC(RUNPLOT):  Parameter DS has value sasuser.houses
MLOGIC(RUNPLOT):  %IF condition %sysprod(graph)=1 is TRUE
MLOGIC(MKTITLE):  Beginning execution.
MLOGIC(MKTITLE):  Parameter PROC has value gplot
MLOGIC(MKTITLE):  Parameter DATA has value sasuser.houses
MLOGIC(MKTITLE):  Ending execution.
MLOGIC(RUNPLOT):  Ending execution.
```

MLOGICNEST System Option

Specifies whether to display the macro nesting information in the MLOGIC output in the SAS log.

Valid in: Configuration fileOPTIONS windowOPTIONS statementSAS invocation

PROC OPTIONS GROUP= MACRO
LOGCONTROL

Type: System option

Default: NOMLOGICNEST

See: "The SAS Log" in Chapter 9 of *SAS Language Reference: Concepts*

Syntax

MLOGICNEST | NOMLOGICNEST

Required Arguments

MLOGICNEST
enables the macro nesting information to be displayed in the MLOGIC output in the SAS log.

NOMLOGICNEST
prevents the macro nesting information from being displayed in the MLOGIC output in the SAS log.

Details

MLOGICNEST enables the macro nesting information to be written to the SAS log in the MLOGIC output.

The setting of MLOGICNEST does not affect the output of any currently executing macro.

The setting of MLOGICNEST does not imply the setting of MLOGIC. You must set both MLOGIC and MLOGICNEST in order for output (with nesting information) to be written to the SAS log.

Example: Using MLOGICNEST System Option

The first example shows both the MLOGIC and MLOGICNEST options being set:

```
%macro outer;
    %put THIS IS OUTER;
    %inner;
%mend outer;
%macro inner;
    %put THIS IS INNER;
    %inrmost;
%mend inner;
%macro inrmost;
    %put THIS IS INRMOST;
%mend;
    options mlogic mlogicnest;
    %outer
```

Here is the MLOGIC output in the SAS log using the MLOGICNEST option:

```
MLOGIC(OUTER):  Beginning execution.
MLOGIC(OUTER):  %PUT THIS IS OUTER
THIS IS OUTER
MLOGIC(OUTER.INNER):  Beginning execution.
MLOGIC(OUTER.INNER): %PUT THIS IS INNER
THIS IS INNER
MLOGIC(OUTER.INNER.INRMOST):  Beginning execution.
MLOGIC(OUTER.INNER.INRMOST):  %PUT THIS IS INRMOST
THIS IS INRMOST
MLOGIC(OUTER.INNER.INRMOST): Ending execution.
```

```
MLOGIC(OUTER.INNER):  Ending execution.
MLOGIC(OUTER):  Ending execution.
```

The second example uses only the NOMLOGICNEST option:

```
%macro outer;
    %put THIS IS OUTER;
    %inner;
%mend outer;
%macro inner;
    %put THIS IS INNER;
    %inrmost;
%mend inner;
%macro inrmost;
    %put THIS IS INRMOST;
%mend;
    options nomlogicnest;
    %outer
```

Here is the output in the SAS log when you use only the NOMLOGICNEST option:

```
MLOGIC(OUTER):  Beginning execution.
MLOGIC(OUTER):  %PUT THIS IS OUTER
THIS IS OUTER
MLOGIC(INNER):  Beginning execution.
MLOGIC(INNER):  %PUT THIS IS INNER
THIS IS INNER
MLOGIC(INRMOST):  Beginning execution.
MLOGIC(INRMOST):  %PUT THIS IS INRMOST
THIS IS INRMOST
MLOGIC(INRMOST):  Ending execution.
MLOGIC(INNER):  Ending execution.
MLOGIC(OUTER):  Ending execution.
```

MPRINT System Option

Specifies whether SAS statements generated by macro execution are traced for debugging.

Valid in: Configuration fileOPTIONS window OPTIONS statementSAS invocation

PROC OPTIONS GROUP= MACRO
LOGCONTROL

Type: System option

Default: NOMPRINT

See: "MFILE System Option" on page 343 and "The SAS Log" in Chapter 9 of *SAS Language Reference: Concepts*

Syntax

MPRINT | NOMPRINT

Required Arguments

MPRINT
displays the SAS statements that are generated by macro execution. The SAS statements are useful for debugging macros.

NOMPRINT
does not display SAS statements that are generated by macro execution.

Details

The MPRINT option displays the text generated by macro execution. Each SAS statement begins a new line. Each line of MPRINT output is identified with the prefix MPRINT(*macro-name*):, to identify the macro that generates the statement. Tokens that are separated by multiple spaces are printed with one intervening space.

You can direct MPRINT output to an external file by also using the MFILE option and assigning the fileref MPRINT to that file. For more information, see "MFILE System Option" on page 343.

Examples

Example 1: Tracing Generation of SAS Statements

In this example, MPRINT traces the SAS statements that are generated when the macros MKTITLE and RUNPLOT execute:

```
%macro mktitle(proc,data);
    title "%upcase(&proc) of %upcase(&data)";
%mend mktitle;
%macro runplot(ds);
   %if %sysprod(graph)=1 %then
      %do;
         %mktitle (gplot,&ds)
         proc gplot data=&ds;
            plot style*price
                / haxis=0 to 150000 by 50000;
         run;
         quit;
      %end;
   %else
      %do;
         %mktitle (plot,&ds)
         proc plot data=&ds;
            plot style*price;
         run;
         quit;
      %end;
%mend runplot;
options mprint;
%runplot(sasuser.houses)
```

When this program executes, this MPRINT output is written to the SAS log:

```
MPRINT(MKTITLE):   TITLE "GPLOT of SASUSER.HOUSES";
MPRINT(RUNPLOT):   PROC GPLOT DATA=SASUSER.HOUSES;
MPRINT(RUNPLOT):   PLOT STYLE*PRICE / HAXIS=0 TO 150000 BY 50000;
```

```
MPRINT(RUNPLOT):   RUN;
MPRINT(RUNPLOT):   QUIT;
```

Example 2: Directing MPRINT Output to an External File

Adding these statements before the macro call in the previous program sends the MPRINT output to the file DEBUGMAC when the SAS session ends.

```
options mfile mprint;
filename mprint 'debugmac';
```

MPRINTNEST System Option

Specifies whether to display the macro nesting information in the MPRINT output in the SAS log.

Valid in: Configuration fileOPTIONS windowOPTIONS statementSAS invocation

PROC OPTIONS GROUP= MACRO

Type: System option

Default: NOMPRINTNEST

Syntax

MPRINTNEST | NOMPRINTNEST

Required Arguments

MPRINTNEST
enables the macro nesting information to be displayed in the MPRINT output in the SAS log.

NOMPRINTNEST
prevents the macro nesting information from being displayed in the MPRINT output in the SAS log.

Details

MPRINTNEST enables the macro nesting information to be written to the SAS log in the MPRINT output. The MPRINTNEST output has no effect on the MPRINT output that is sent to an external file. For more information, see MFILE System Option.

The setting of MPRINTNEST does not imply the setting of MPRINT. You must set both MPRINT and MPRINTNEST in order for output (with the nesting information) to be written to the SAS log.

Example: Using MPRINTNEST System Option

The following example uses the MPRINT and MPRINTNEST options:

```
%macro outer;
data _null_;
     %inner
run;
%mend outer;
%macro inner;
```

```
    put %inrmost;
%mend inner;
%macro inrmost;
    'This is the text of the PUT statement'
%mend inrmost;
    options mprint mprintnest;
    %outer
```

Here is the output written to the SAS log using both the MPRINT option and the MPRINTNEST option:

```
MPRINT(OUTER):   data _null_;
MPRINT(OUTER.INNER):   put
MPRINT(OUTER.INNER.INRMOST):   'This is the text of the PUT statement'
MPRINT(OUTER.INNER):  ;
MPRINT(OUTER):   run;
This is the text of the PUT statement
NOTE: DATA statement used (Total process time):
      real time           0.10 seconds
      cpu time            0.06 seconds
```

Here is an example that uses the NOMPRINTNEST option:

```
%macro outer;
    data _null_;
    %inner
run;
%mend outer;
%macro inner;
    put %inrmost;
%mend inner;
%macro inrmost;
    'This is the text of the PUT statement'
%mend inrmost;
    options nomprintnest;
    %outer
```

Here is the output written to the SAS log using the NOMPRINTNEST option:

```
MPRINT(OUTER):   data _null_;
MPRINT(INNER):   put
MPRINT(INRMOST):   'This is the text of the PUT statement'
MPRINT(INNER):  ;
MPRINT(OUTER):   run;
This is the text of the PUT statement
NOTE: DATA statement used (Total process time):
      real time           0.00 seconds
      cpu time            0.01 seconds
```

MRECALL System Option

Specifies whether autocall libraries are searched for a member that was not found during an earlier search.

Valid in: Configuration fileOPTIONS windowOPTIONS statementSAS invocation

PROC OPTIONS GROUP= MACRO

Type: System option

Default: NOMRECALL

Syntax

MRECALL | NOMRECALL

Required Arguments

MRECALL
searches the autocall libraries for an undefined macro name each time an attempt is made to invoke the macro. It is inefficient to search the autocall libraries repeatedly for an undefined macro. Generally, use this option when you are developing or debugging programs that call autocall macros.

NOMRECALL
searches the autocall libraries only once for a requested macro name.

Details

Use the MRECALL option primarily for

- developing systems that require macros in autocall libraries.
- recovering from errors caused by an autocall to a macro that is in an unavailable library. Use MRECALL to call the macro again after making the library available. In general, do not use MRECALL unless you are developing or debugging autocall macros.

MREPLACE System Option

Specifies whether to enable existing macros to be redefined.

Valid in: Configuration fileOPTIONS windowOPTIONS statementSAS invocation

PROC OPTIONS GROUP= MACRO

Type: System option

Default: MREPLACE

Syntax

MREPLACE | NOMREPLACE

Required Arguments

MREPLACE
enables you to redefine existing macro definitions that are stored in a catalog in the WORK library.

NOMREPLACE
prevents you from redefining existing macro definitions that are stored in a catalog in the WORK library.

Details

The MREPLACE system option enables you to overwrite existing macros if the names are the same.

The NOMREPLACE system option prevents you from overwriting a macro even if a macro with the same name has already been compiled.

MSTORED System Option

Specifies whether the macro facility searches a specific catalog for a stored compiled macro.

Valid in: Configuration fileOPTIONS windowOPTIONS statementSAS invocation

PROC OPTIONS GROUP= MACRO

Type: System option

Default: NOMSTORED

Syntax

MSTORED | NOMSTORED

Required Arguments

MSTORED
searches for stored compiled macros in a catalog in the SAS library referenced by the SASMSTORE= option.

NOMSTORED
does not search for compiled macros.

Details

Regardless of the setting of MSTORED, the macro facility first searches for macros compiled in the current SAS session. If the MSTORED option is in effect, the macro facility next searches the libraries containing compiled stored macros. If the MAUTOSOURCE option is in effect, the macro facility next searches the autocall libraries. Then, the macro facility searches the SASMACR catalog in the SASHELP library.

MSYMTABMAX= System Option

Specifies the maximum amount of memory available to the macro variable symbol table or tables.

Valid in: Configuration fileOPTIONS windowOPTIONS statement SAS invocation

PROC OPTIONS GROUP= MACRO

Type: System option

Syntax

MSYMTABMAX= *n* | *n*K | *n*M | *n*G | MAX

Required Arguments

n
specifies the maximum memory available in bytes.

***n*K**
specifies the maximum memory available in kilobytes.

***n*M**
specifies the maximum memory available in megabytes.

***n*G**
specifies the maximum memory available in gigabytes.

MAX
specifies the maximum memory of 65534.

Details

Once the maximum value is reached, additional macro variables are written out to disk.

The value that you specify with the MSYMTABMAX= system option can range from 0 to the largest nonnegative integer representable on your operating environment. The default values are host dependent. A value of 0 causes all macro symbol tables to be written to disk.

The value of MSYMTABMAX= can affect system performance. If this option is set too low and the application frequently reaches the specified memory limit, then disk I/O increases. If this option is set too high (on some operating environments) and the application frequently reaches the specified memory limit, then less memory is available for the application, and CPU usage increases. Before you specify the value for production jobs, run tests to determine the optimum value.

MVARSIZE= System Option

Specifies the maximum size for macro variable values that are stored in memory.

Valid in: Configuration fileOPTIONS windowOPTIONS statement SAS invocation

PROC OPTIONS GROUP= MACRO

Type: System option

Syntax

MVARSIZE= *n* | *n*K | *n*M | *n*G | MAX

Required Arguments

n
specifies the maximum memory available in bytes.

***n*K**
specifies the maximum memory available in kilobytes.

***n*M**
specifies the maximum memory available in megabytes.

***n*G**
specifies the maximum memory available in gigabytes.

MAX
specifies the maximum memory of 65534.

Details

If the memory required for a macro variable value is larger than the MVARSIZE= value, the variable is written to a temporary catalog on disk. The macro variable name is used as the member name, and all members have the type MSYMTAB.

The value that you specify with the MVARSIZE= system option can range from 0 to 65534. A value of 0 causes all macro variable values to be written to disk.

The value of MVARSIZE= can affect system performance. If this option is set too low and the application frequently creates macro variables larger than the limit, then disk I/O increases. Before you specify the value for production jobs, run tests to determine the optimum value.

Note: The MVARSIZE= option has no affect on the maximum length of the value of the macro variable. For more information, see "Macro Variables" on page 19.

SASAUTOS= System Option

Specifies the location of one or more autocall libraries.

Valid in: Configuration fileOPTIONS windowOPTIONS statementSAS invocation

PROC OPTIONS GROUP= ENVFILES
MACRO

Type: System option

Syntax

SASAUTOS= *library-specification* |
(*library-specification-1 . . . , library-specification-n*)

Required Arguments

library-specification
identifies a location that contains library members that contain a SAS macro definition. A location can be a SAS fileref or a host-specific location name enclosed in quotation marks. Each member contains a SAS macro definition.

(*library-specification-1 . . . , library-specification-n*)
identifies two or more locations that contain library members that contain a SAS macro definition. A location can be a SAS fileref or a host-specific location name enclosed in quotation marks. When you specify two or more autocall libraries, enclose the specifications in parentheses and separate them with either a comma or a blank space.

Details

When SAS searches for an autocall macro definition, it opens and searches each location in the same order that it is specified in the SASAUTOS option. If SAS cannot open any specified location, it generates a warning message and sets the NOMAUTOSOURCE system option on. To use the autocall facility again in the same SAS session, you must specify the MAUTOSOURCE option again.

Operating Environment Information
You specify a source library by using a fileref or by enclosing the host-specific location name in quotation marks. A valid library specification and its syntax are host specific. Although the syntax is generally consistent with the command-line syntax of your operating environment, it might include additional or alternate punctuation. For details, see the SAS documentation for your operating environment.

z/OS Specifics
You can use the APPEND or INSERT system options to add additional *library-specification*. For details, see the documentation for the APPEND and INSERT system options under UNIX and z/OS.

SASMSTORE= System Option

Identifies the libref of a SAS library with a catalog that contains, or will contain, stored compiled SAS macros.

Valid in: Configuration file, OPTIONS window, OPTIONS statement, SAS invocation

PROC OPTIONS GROUP= MACRO

Type: System option

Syntax

SASMSTORE=*libref*

Required Argument

libref
specifies the libref of a SAS library that contains, or will contain, a catalog of stored compiled SAS macros. This libref cannot be WORK.

SERROR System Option

Specifies whether the macro processor issues a warning message when a macro variable reference does not match a macro variable.

Valid in: Configuration fileOPTIONS window OPTIONS statement SAS invocation

PROC OPTIONS GROUP= MACRO

Type: System option

Alias: SERR | NOSERR

Default: SERROR

Syntax

SERROR | NOSERROR

Required Arguments

SERROR
issues a warning message when the macro processor cannot match a macro variable reference to an existing macro variable.

NOSERROR
issues no warning messages when the macro processor cannot match a macro variable reference to an existing macro variable.

Details

Several conditions can occur that prevent a macro variable reference from resolving. These conditions appear when one or more of the following is true:

- the name in a macro variable reference is misspelled.
- the variable is referenced before being defined.
- the program contains an ampersand (&) followed by a string, without intervening blanks between the ampersand and the string. For example:

```
if x&y then do;
if buyer="Smith&Jones, Inc." then do;
```

If your program uses a text string containing ampersands and you want to suppress the warnings, specify NOSERROR.

SYMBOLGEN System Option

Specifies whether the results of resolving macro variable references are written to the SAS log for debugging.

Valid in: Configuration fileOPTIONS window OPTIONS statementSAS invocation

PROC OPTIONS GROUP= MACRO
LOGCONTROL

Type: System option

Alias: SGEN | NOSGEN

Default: NOSYMBOLGEN

See: "The SAS Log" in Chapter 9 of *SAS Language Reference: Concepts*

Syntax

SYMBOLGEN | NOSYMBOLGEN

Required Arguments

SYMBOLGEN
displays the results of resolving macro variable references. This option is useful for debugging.

NOSYMBOLGEN
does not display results of resolving macro variable references.

Details

SYMBOLGEN displays the results in this form:

```
SYMBOLGEN: Macro variable name resolves to value
```

SYMBOLGEN also indicates when a double ampersand (`&&`) resolves to a single ampersand (`&`).

Example: Tracing Resolution of Macro Variable References

In this example, SYMBOLGEN traces the resolution of macro variable references when the macros MKTITLE and RUNPLOT execute:

```
%macro mktitle(proc,data);
    title "%upcase(&proc) of %upcase(&data)";
%mend mktitle;
%macro runplot(ds);
   %if %sysprod(graph)=1 %then
      %do;
         %mktitle (gplot,&ds)
         proc gplot data=&ds;
            plot style*price
                / haxis=0 to 150000 by 50000;
         run;
         quit;
      %end;
   %else
      %do;
         %mktitle (plot,&ds)
         proc plot data=&ds;
            plot style*price;
         run;
         quit;
      %end;
%mend runplot;
%runplot(sasuser.houses)
```

When this program executes, this SYMBOLGEN output is written to the SAS log:

```
SYMBOLGEN:  Macro variable DS resolves to sasuser.houses
SYMBOLGEN:  Macro variable PROC resolves to gplot
SYMBOLGEN:  Macro variable DATA resolves to sasuser.houses
SYMBOLGEN:  Macro variable DS resolves to sasuser.houses
```

SYSPARM= System Option

Specifies a character string that can be passed to SAS programs.

Valid in: Configuration fileOPTIONS windowOPTIONS statementSAS invocation

Type: System option

Syntax

SYSPARM='*character-string*'

Required Argument

character-string
is a character string, enclosed in quotation marks, with a maximum length of 200.

Details

The character string specified can be accessed in a SAS DATA step by the SYSPARM() function or anywhere in a SAS program by using the automatic macro variable reference &SYSPARM.

Operating Environment Information
The syntax shown here applies to the OPTIONS statement. At invocation, on the command line, or in a configuration file, the syntax is host specific. For details, see the SAS documentation for your operating environment.

Example: Passing a User Identification to a Program

This example uses the SYSPARM option to pass a user identification to a program.

```
options sysparm='usr1';
data a;
   length z $100;
   if sysparm()='usr1' then z="&sysparm";
run;
```

Part 3

Appendices

Appendix 1
Reserved Words in the Macro Facility

Macro Facility Word Rules

The following rules apply to the macro facility.

- Do not use a reserved word as the name of a macro, a macro variable, or a macro label. Reserved words include words reserved by both the macro facility and the operating environment. When a macro name is a macro facility reserved word, the macro processor issues a warning, and the macro is neither compiled nor available for execution. The macro facility reserves the words listed under "Reserved Words" on page 363 for internal use.
- Do not prefix the name of a macro language element with SYS because SAS reserves the SYS prefix for the names of macro language elements supplied with SAS software.
- Do not prefix macro variables names with SYS, AF, or DMS in order to avoid macro name conflicts.

Reserved Words

The following table lists the reserved words for the macro facility.

Table A1.1 *Macro Facility Reserved Words*

ABEND	END	LENGTH	QKUPCASE	SYSEVALF
ABORT	EVAL	LET	QSCAN	SYSEXEC
ACT	FILE	LIST	QSUBSTR	SYSFUNC
ACTIVATE	GLOBAL	LISTM	QSYSFUNC	SYSGET

BQUOTE	GO	LOCAL	QUOTE	SYSRPUT
BY	GOTO	MACRO	QUPCASE	THEN
CLEAR	IF	MEND	RESOLVE	TO
CLOSE	INC	PAUSE	RETURN	TSO
CMS	INCLUDE	NRSTR	RUN	UNQUOTE
COMANDR	INDEX	ON	SAVE	UNSTR
COPY	INFILE	OPEN	SCAN	UNTIL
DEACT	INPUT	PUT	STOP	UPCASE
DEL	KCMPRES	NRBQUOTE	STR	WHILE
DELETE	KINDEX	NRQUOTE	SYSCALL	WINDOW
DISPLAY	KLEFT	METASYM	SUBSTR	
DMIDSPLY	KLENGTH	QKCMPRES	SUPERQ	
DMISPLIT	KSCAN	QKLEFT	SYMDEL	
DO	KSUBSTR	QKSCAN	SYMEXIST	
EDIT	KTRIM	QKSUBSTR	SYMGLOBL	
ELSE	KUPCASE	QKTRIM	SYMLOCAL	

Appendix 2
SAS Tokens

SAS Tokens

When SAS processes a program, a component called the word scanner reads the program, character by character, and groups the characters into words. These words are referred to as tokens.

List of Tokens

SAS recognizes four general types of tokens:

Literal
: One or more characters enclosed in single or double quotation marks. Examples of literals include the following:

Table A2.1 *Literals*

'CARY'	"2008"
'Dr. Kemple-Long'	'<entry align="center">'

Name
: One or more characters beginning with a letter or an underscore. Other characters can be letters, underscores, and digits.

Table A2.2 *Names*

data	_test	linesleft
f25	univariate	otherwise

year_2008	descending

Number

A numeric value. Number tokens include the following:

- integers. Integers are numbers that do not contain a decimal point or an exponent. Examples of integers include 1, 72, and 5000. SAS date, time, and datetime constants such as '24AUG2008'D are integers, as are hexadecimal constants such as 0C4X.
- real (floating-point) numbers. Floating-point numbers contain a decimal point or an exponent. Examples include numbers such as 2.35, 5., 2.3E1, and 5.4E− 1.

Special character

Any character that is not a letter, number, or underscore. The following characters are some special characters:

= + − % & ; ( ) #

The maximum length of any type of token is 32,767 characters. A token ends when the tokenizer encounters one of the following situations:

- the beginning of a new token.
- a blank after a name or number token.
- in a literal token, a quotation mark of the same type that started the token. There is an exception. A quotation mark followed by a quotation mark of the same type is interpreted as a single quotation mark that becomes part of the literal token. For example, in `'Mary''s'`, the fourth quotation mark terminates the literal token. The second and third quotation marks are interpreted as a single character, which is included in the literal token.

Appendix 3

Syntax for Selected Functions Used with the %SYSFUNC Function

Summary Descriptions and Syntax

This appendix provides summary descriptions and syntax for selected functions that can be used with the %SYSFUNC function.

Functions and Arguments for %SYSFUNC

The following table shows the syntax for selected functions that can be used with the %SYSFUNC function. This is not a complete list of the functions that can be used with %SYSFUNC. For a list of functions that cannot be used with %SYSFUNC, see Table 17.2 on page 269.

Table A3.1 *Functions and Arguments for %SYSFUNC*

Function	Description and Syntax
ATTRC	Returns the value of a character attribute for a SAS data set. **%SYSFUNC(ATTRC(*data-set-id,attr-name*))**
ATTRN	Returns the value of a numeric attribute for specified SAS data set. **%SYSFUNC(ATTRN(*data-set_id,attr-name*))**
CEXIST	Verifies the existence of a SAS catalog or SAS catalog entry. **%SYSFUNC(CEXIST(*entry <, U>*))**
CLOSE	Closes a SAS data set.**%SYSFUNC(CLOSE(*data-set-id*))**
CUROBS	Returns the number of the current observation. **%SYSFUNC(CUROBS(*data-set-id*))**
DCLOSE	Closes a directory.**%SYSFUNC(DCLOSE(*directory-id*))**

Function	Description and Syntax
DINFO	Returns specified information items for a directory. %SYSFUNC(DINFO(*directory-id,info-items*))
DNUM	Returns the number of members in a directory. %SYSFUNC(DNUM(*directory-id*))
DOPEN	Opens a directory.%SYSFUNC(DOPEN(*fileref*))
DOPTNAME	Returns a specified directory attribute. %SYSFUNC(DOPTNAME(*directory-id,nval*))
DOPTNUM	Returns the number of information items available for a directory. %SYSFUNC(DOPTNUM(*directory-id*))
DREAD	Returns the name of a directory member. %SYSFUNC(DREAD(*directory-id,nval*))
DROPNOTE	Deletes a note marker from a SAS data set or an external file. %SYSFUNC(DROPNOTE(*data-set-id* | *file-id,note-id*))
DSNAME	Returns the data set name associated with a data set identifier. %SYSFUNC(DSNAME(<*data-set-id*>))
EXIST	Verifies the existence of a SAS library member. %SYSFUNC(EXIST(*member-name*<,*member-type*>))
FAPPEND	Appends a record to the end of an external file. %SYSFUNC(FAPPEND(*file-id*<,*cc*>))
FCLOSE	Closes an external file, directory, or directory member. %SYSFUNC(FCLOSE(*file-id*))
FCOL	Returns the current column position in the File Data Buffer (FDB) %SYSFUNC(FCOL(*file-id*))
FDELETE	Deletes an external file. %SYSFUNC(FDELETE(*fileref*))
FETCH	Reads the next nondeleted observation from a SAS data set into the Data Set Data Vector (DDV). %SYSFUNC(FETCH(*data-set-id*<,NOSET>))
FETCHOBS	Reads a specified observation from a SAS data set into the DDV. %SYSFUNC(FETCHOBS(*data-set-id,obs-number*<,*options*>))
FEXIST	Verifies the existence of an external file associated with a fileref. %SYSFUNC(FEXIST(*fileref*))
FGET	Copies data from the FDB. %SYSFUNC(FGET(*file-id,cval*<,*length*>))

Function	Description and Syntax
FILEEXIST	Verifies the existence of an external file by its physical name. %SYSFUNC(FILEEXIST(*file-name*))
FILENAME	Assigns or deassigns a fileref for an external file, directory, or output device. %SYSFUNC(FILENAME(*fileref,file-name<,device<,host-options<,dir-ref>>>*))
FILEREF	Verifies that a fileref has been assigned for the current SAS session. %SYSFUNC(FILEREF(*fileref*))
FINFO	Returns a specified information item for a file. %SYSFUNC(FINFO(*file-id,info-item*))
FNOTE	Identifies the last record that was read. %SYSFUNC(FNOTE(*file-id*))
FOPEN	Opens an external file. %SYSFUNC(FOPEN(*fileref<,open-mode<,record-length<,record-format>>>*))
FOPTNAME	Returns the name of an information item for an external file. %SYSFUNC(FOPTNAME(*file-id,nval*))
FOPTNUM	Returns the number of information items available for an external file. %SYSFUNC(FOPTNUM(*file-id*))
FPOINT	Positions the read pointer on the next record to be read. %SYSFUNC(FPOINT(*file-id,note-id*))
FPOS	Sets the position of the column pointer in the FDB. %SYSFUNC(FPOS(*file-id,nval*))
FPUT	Moves data to the FDB of an external file starting at the current column position. %SYSFUNC(FPUT(*file-id,cval*))
FREAD	Reads a record from an external file into the FDB. %SYSFUNC(FREAD(*file-id*))
FREWIND	Positions the file pointer at the first record. %SYSFUNC(FREWIND(*file-id*))
FRLEN	Returns the size of the last record read, or the current record size for a file opened for output. %SYSFUNC(FRLEN(*file-id*))
FSEP	Sets the token delimiters for the FGET function. %SYSFUNC(FSEP(*file-id,cval*))
FWRITE	Writes a record to an external file. %SYSFUNC(FWRITE(*file-id<,cc>*))

Function	Description and Syntax
GETOPTION	Returns the value of a SAS system or graphics option. %SYSFUNC(GETOPTION(*option-name<,reporting-options<,...>>*))
GETVARC	Assigns the value of a SAS data set variable to a character DATA step or macro variable. %SYSFUNC(GETVARC(*data-set-id,var-num*))
GETVARN	Assigns the value of a SAS data set variable to a numeric DATA step or macro variable. %SYSFUNC(GETVARN(*data-set-id,var-num*))
LIBNAME	Assigns or deassigns a libref for a SAS library. %SYSFUNC(LIBNAME(*libref<,SAS-data-library<,engine<,options>>>*))
LIBREF	Verifies that a libref has been assigned. %SYSFUNC(LIBREF(*libref*))
MOPEN	Opens a directory member file. %SYSFUNC(MOPEN(*directory-id,member-name<open-mode<,record-length<,record-format>>>*))
NOTE	Returns an observation ID for current observation of a SAS data set. %SYSFUNC(NOTE(*data-set-id*))
OPEN	Opens a SAS data file. %SYSFUNC(OPEN(*<data-file-name<,mode>>*))
PATHNAME	Returns the physical name of a SAS library or an external file. %SYSFUNC(PATHNAME(*fileref*))
POINT	Locates an observation identified by the NOTE function. %SYSFUNC(POINT(*data-set-id,note-id*))
REWIND	Positions the data set pointer to the beginning of a SAS data set. %SYSFUNC(REWIND(*data-set-id*))
SPEDIS	Returns a number for the operation required to change an incorrect keyword in a WHERE clause to a correct keyword. %SYSFUNC(SPEDIS(*query,keyword*))
SYSGET	Returns the value of the specified host environment variable. %SYSFUNC(sysget(*host-variable*))
SYSMSG	Returns the error or warning message produced by the last function that attempted to access a data set or external file. %SYSFUNC(SYSMSG())
SYSRC	Returns the system error number or exit status of the entry most recently called. %SYSFUNC(SYSRC())

Function	Description and Syntax
VARFMT	Returns the format assigned to a data set variable. `%SYSFUNC(VARFMT(data-set-id, var-num))`
VARINFMT	Returns the informat assigned to a data set variable. `%SYSFUNC(VARINFMT(data-set-id, var-num))`
VARLABEL	Returns the label assigned to a data set variable. `%SYSFUNC(VARLABEL(data-set-id, var-num))`
VARLEN	Returns the length of a data set variable. `%SYSFUNC(VARLEN(data-set-id, var-num))`
VARNAME	Returns the name of a data set variable. `%SYSFUNC(VARNAME(data-set-id, var-num))`
VARNUM	Returns the number of a data set variable. `%SYSFUNC(VARNUM(data-set-id, var-name))`
VARTYPE	Returns the data type of a data set variable. `%SYSFUNC(VARTYPE(data-set-id, var-num))`

Glossary

arithmetic expression
a type of macro expression that consists of a sequence of arithmetic operators and operands. An arithmetic expression returns a numeric value when it is executed.

autocall facility
a feature of SAS that enables you to store the source statements that define a macro and to invoke the macro as needed, without having to include the definition in your program.

autocall macro
a macro whose uncompiled source code and text are stored in an autocall macro library. Unlike a stored compiled macro, an autocall macro is compiled before execution the first time it is called.

command-style macro
a macro that is defined with the CMD option in the %MACRO statement.

constant text
the character strings that are stored as part of a macro or as a macro variable's value in open code, from which the macro processor generates text to be used as SAS statements, display manager commands, or other macro program statements. Constant text is also called model text.

dummy macro
a macro that the macro processor compiles but does not store.

global macro variable
a macro variable that can be referenced in either global or local scope in a SAS program, except where there is a local macro variable that has the same name. A global macro variable exists until the end of the session or program.

global scope
in SAS macro programming, indicates broad context boundaries for referencing global macro variables; that is, anywhere within the current SAS session or SAS batch program.

input stack
the most recently read line of input from a SAS program and any text generated by the macro processor that is awaiting processing by the word scanner.

keyword parameter
a type of macro parameter that is identified by its name, followed by an equals sign. Multiple keyword parameters can be provided in any order, and must follow any positional parameters.

local macro variable
a macro variable that is available only within the macro in which it was created and within macros that are invoked from within that macro. A local macro variable ceases to exist when the macro that created it stops executing.

local scope
in SAS macro programming, indicates narrowed context boundaries for referencing local macro variables; that is, limited to the current macro.

logical expression
a type of macro expression that consists of a sequence of logical operators and operands. A logical expression returns a value of either true or false when it is executed.

macro
a SAS catalog entry that contains a group of compiled program statements and stored text.

macro call
within a SAS program, a statement that invokes (or calls) a stored compiled macro program. You use the syntax %<user-sup-val>macro-name</user-sup-val>; to call a macro.

macro compilation
the process of converting a macro definition from the statements that you enter to a form that is ready for the macro processor to execute. The compiled macro is then stored for later use in the SAS program or session.

macro execution
the process of following the instructions that are given by compiled macro program statements in order to generate text, to write messages to the SAS log, to accept input, to create or change the values of macro variables, or to perform other activities. The generated text can be a SAS statement, a SAS command, or another macro program statement.

macro expression
any valid combination of symbols that returns a value when it is executed. The three types of macro expressions are text, logical, and arithmetic. A text expression generates text when it is resolved (executed) and can consist of any combination of text, macro variables, macro functions, and macro calls. A logical expression consists of logical operators and operands and returns a value of either true or false. An arithmetic expression consists of arithmetic operators and operands and returns a numeric value.

macro facility
a component of Base SAS software that you can use for extending and customizing SAS programs and for reducing the amount of text that must be entered in order to perform common tasks. The macro facility consists of the macro processor and the macro programming language.

macro function
a function that is defined by the macro facility. Each macro function processes one or more arguments and produces a result.

macro invocation
another term for macro call.

macro language
the programming language that is used to communicate with the macro processor.

macro parameter
a local macro variable that is defined within parentheses in a %MACRO statement. You supply values to a macro parameter when you invoke a macro.

macro processor
the component of SAS software that compiles and executes macros and macro program statements.

macro quoting
a function that tells the macro processor to interpret special characters and mnemonics as text rather than as part of the macro language.

macro variable
a variable that is part of the SAS macro programming language. The value of a macro variable is a string that remains constant until you change it. Macro variables are sometimes referred to as symbolic variables.

macro variable reference
the name of a macro variable, preceded by an ampersand (&<user-sup-val>name</user-sup-val>). The macro processor replaces the macro variable reference with the value of the specified macro variable.

model text
another term for constant text.

name-style macro
a macro that is named and defined with the %MACRO statement.

null value
in the SAS macro language, a value that consists of zero characters.

open code
the part of a SAS program that is outside any macro definition.

positional parameter
a type of macro parameter that is named (using comma delimiters) in the %MACRO statement at invocation, and is defined in the corresponding position (again using comma delimiters) in the macro execution statement.

quoting
the process that causes the macro processor to read certain items as text rather than as symbols in the macro language. Quoting is also called removing the significance of an item and treating an item as text.

quoting function
a macro language function that performs quoting on its argument.

reserved word
a name that is reserved for use by an internal component of a software application and which therefore cannot be assigned by a user of that application to any type of data object.

returned value
a character string that is the result of the execution of a macro function.

SAS compilation
the process of converting statements in the SAS language from the form in which you enter them to a form that is ready for SAS to use.

SAS variable
a column in a SAS data set or in a SAS data view. The data values for each variable describe a single characteristic for all observations (rows).

scope
in programming, the enclosing context for associated values and expressions. In SAS macro programming, the scope can be either global or local, which will determine how values are assigned to a macro variable and how the macro processor resolves references to it.

session compiled macro
a macro that the macro processor compiles and stores in a SAS catalog in the WORK library. These macros exist only during the current SAS session. Unlike stored compiled macros, session compiled macros cannot be called in any other SAS session.

statement-style macro
a macro that is defined with the STMT option in the %MACRO statement.

stored compiled macro
a macro program that was compiled in a previous session and which was stored in a permanent directory. Unlike session compiled macros, stored compiled macros can be called in any SAS program.

string
in the SAS macro language, any group of consecutive characters.

symbol table
the area in which the macro processor stores all macro variables and macro statement labels for a particular scope.

symbolic substitution
the process of resolving a macro variable reference (&<user-sup-val>variable-name</user-sup-val>) to its value.

symbolic variable
another term for macro variable.

text expression
a type of macro expression that generates text when it is resolved (executed). The text expression can include any combination of text, macro variables, macro functions, and macro calls.

token

the unit into which the SAS language or the macro language divides input in order to enable SAS to process that input. Tokens (also called words) include items that look like English words (such as variable names) as well as items that do not (such as mathematical operators and semicolons).

tokenizer

the part of the word scanner that divides input into tokens (also called words).

unquoting

the process of restoring the meaning of a quoted item.

word

another term for token.

word scanner

the component of SAS that examines all tokens (words) in a SAS program and moves the tokens to the correct component of SAS for processing.

Index

N

S

T

U

V

W

Your Turn

We welcome your feedback.

- If you have comments about this book, please send them to **yourturn@sas.com**. Include the full title and page numbers (if applicable).

- If you have comments about the software, please send them to **suggest@sas.com**.

